AF581766

Deep Learning for Facial Informatics

Deep Learning for Facial Informatics

Editors

Gee-Sern Jison Hsu
Radu Timofte

MDPI • Basel • Beijing • Wuhan • Barcelona • Belgrade • Manchester • Tokyo • Cluj • Tianjin

Editors
Gee-Sern Jison Hsu
Artificial Vision Laboratory
Department of Mechanical Engineering
National Taiwan University of Science and Technology
Taiwan

Radu Timofte
Computer Vision Laboratory
ETH Zurich
Switzerland

Editorial Office
MDPI
St. Alban-Anlage 66
4052 Basel, Switzerland

This is a reprint of articles from the Special Issue published online in the open access journal *Symmetry* (ISSN 2073-8994) (available at: https://www.mdpi.com/journal/symmetry/special_issues/Deep_Learning_Face_Informatics).

For citation purposes, cite each article independently as indicated on the article page online and as indicated below:

LastName, A.A.; LastName, B.B.; LastName, C.C. Article Title. *Journal Name* **Year**, *Article Number*, Page Range.

ISBN 978-3-03936-964-5 (Hbk)
ISBN 978-3-03936-965-2 (PDF)

Contents

About the Editors

Gee-Sern Jison Hsu completed his dual MS degree in electrical and mechanical engineering and his Ph.D. in mechanical engineering at the University of Michigan, Ann Arbor, in 1993 and 1995, respectively. From 1995 to 1996, he was a post-doctoral fellow at the University of Michigan. From 1997 to 2000, he was a senior research staff member at the National University of Singapore. In 2001, he joined Penpower Technology, where he led research on face recognition and intelligent video surveillance. His team at Penpower Technology were recipients of the Best Innovation and Best Product Awards at the SecuTech Expo for three consecutive years. In 2007, he joined the Department of Mechanical Engineering, National Taiwan University of Science and Technology (NTUST), where he is now an associate professor. His research interests include deep learning, computer vision and pattern recognition. He serves as a reviewer for major journals, including TIP, TIFS, TCSVT, PR, CVIU and TNSRE; and major conferences, e.g., ECCV and ICME. He received best paper awards in ICMT 2011, CVGIP 2013, CVPRW 2014, ARIS 2017 and CVGIP 2018. He is a senior member of IEEE and IAPR.

Radu Timofte is a lecturer and research group leader at the Computer Vision Laboratory, ETH Zurich, Switzerland. He obtained a Ph.D. in Electrical Engineering at KU Leuven, Belgium, in 2013; MSc at the Univ. of Eastern Finland in 2007; and Dipl. Eng. at the Technical Univ. of Iasi, Romania, in 2006. He serves as a reviewer for top journals (such as TPAMI, TIP, IJCV, TNNLS, TCSVT, CVIU, PR) and conferences (ICCV, CVPR, ECCV, NeurIPS), and is an associate editor for *Elsevier CVIU journal* and, starting in 2020, for *IEEE Trans. PAMI* and for *SIAM Journal on Imaging Sciences*. He served as an area chair for ACCV 2018, ICCV 2019 and ECCV 2020, and as a senior PC member for IJCAI 2019 and 2020. He received a NIPS 2017 best reviewer award. His work received the best student paper award at BMVC 2019, a best scientific paper award at ICPR 2012, the best paper award at CVVT workshop (ECCV 2012), the best paper award at ChaLearn LAP workshop (ICCV 2015), the best scientific poster award at EOS 2017, the honorable mention award at FG 2017, and his team won a number of challenges, including traffic sign detection (IJCNN 2013), apparent age estimation (ICCV 2015) and real world super-resolution (ICCV 2019). He is a co-founder of Merantix and co-organizer of NTIRE, CLIC, AIM and PIRM events. His current research interests include sparse and collaborative representations, deep learning, optical flow, image/video compression, restoration and enhancement.

Preface to "Deep Learning for Facial Informatics"

Deep learning has been revolutionizing many fields in computer vision, and facial informatics is one of the major fields. Novel approaches and performance breakthroughs are often reported on existing benchmarks. As the performances on existing benchmarks are close to saturation, larger and more challenging databases are being made and considered as new benchmarks, further pushing the advancement of the technologies. Considering face recognition, for example, the VGG-Face2 and Dual-Agent GAN report nearly perfect and better-than-human performances on the IARPA Janus Benchmark A (IJB-A) benchmark. More challenging benchmarks, e.g., the IARPA Janus Benchmark C (IJB-C), QMUL-SurvFace and MegaFace, are accepted as new standards for evaluating the performance of a new approach. Such an evolution is also seen in other branches of face informatics. In this Special Issue, we have selected papers that report the latest progresses made in the following topics:

1. Face Liveness Detection
2. Emotion Classification
3. Facial Age Estimation
4. Facial Landmark Detection

We would like to thank all of the authors who have submitted their work to this Special Issue, and the reviewers who have contributed their time for the review. We wish the readers to be able to gain some new perspectives of this interesting field. We would also like to thank MDPI for publishing this Special Issue.

Gee-Sern Jison Hsu, Radu Timofte
Editors

Article

Face Liveness Detection Using Thermal Face-CNN with External Knowledge

Jongwoo Seo [1] and In-Jeong Chung [2,*]

1 Department of Computer and Information Science, Korea University, Sejong Campus, Sejong City 30019, Korea; sjw007s@korea.ac.kr
2 Department of Computer Convergence Software, Korea University, Sejong Campus, Sejong City 30019, Korea
* Correspondence: chung@korea.ac.kr

Received: 1 January 2019; Accepted: 6 March 2019; Published: 10 March 2019

Abstract: Face liveness detection is important for ensuring security. However, because faces are shown in photographs or on a display, it is difficult to detect the real face using the features of the face shape. In this paper, we propose a thermal face-convolutional neural network (Thermal Face-CNN) that knows the external knowledge regarding the fact that the real face temperature of the real person is 36~37 degrees on average. First, we compared the red, green, and blue (RGB) image with the thermal image to identify the data suitable for face liveness detection using a multi-layer neural network (MLP), convolutional neural network (CNN), and C-support vector machine (C-SVM). Next, we compared the performance of the algorithms and the newly proposed Thermal Face-CNN in a thermal image dataset. The experiment results show that the thermal image is more suitable than the RGB image for face liveness detection. Further, we also found that Thermal Face-CNN performs better than CNN, MLP, and C-SVM when the precision is slightly more crucial than recall through F-measure.

Keywords: face liveness detection; convolutional neural network; thermal image; external knowledge

---

1. Introduction

Face liveness detection in indoor residential environments is an important technique for delivering security information, such as in the case of unlocking a mobile device using a face recognition system. For example, in order to allow access to only one specific person, that person's unique information, such as their face, can be used to unlock security measures. However, because the printed face photograph and face from the display can sufficiently generate the unique information of the face, the reliability of the security is reduced. Therefore, there is a need to provide more secure security by using face liveness detection, in which thermal images are distinguishable between the real face and the fake face through the heat distribution existing in the face of the real person.

In this paper, we first quantitatively identify a more suitable image for face liveness detection using both the RGB image and the thermal image. The same algorithms were applied to the RGB and thermal image datasets for the comparison. A multi-layer neural network (MLP) [1], convolutional neural network (CNN) [2], and C-support vector machine (C-SVM) [3] with a smooth hyperplane were used for the comparison. In addition, we compared the performance of the existing algorithms with thermal face-convolutional neural network (Thermal Face-CNN) proposed in this paper. Thermal Face-CNN is an algorithm with external knowledge about the temperature values that are found in a real face.

We have collected thermal images because there are many RGB image datasets for face liveness detection but few or no thermal image datasets available. We obtained RGB and thermal images of the same scene in order to evaluate how these thermal images improve performance over RGB images. Accuracy [4], recall [4], and precision [4] were mainly obtained on both the RGB and thermal image datasets.

The experimental results show that the best-performing CNN performance has an accuracy of 0.6898, a recall of 0.5752, and a precision of 0.7342 on the RGB image dataset, while it has an accuracy of 0.8367, a recall of 0.7876, and a precision of 0.8476 on the thermal image dataset. Therefore, it has been shown that the thermal image is more effective in face liveness detection than the RGB image. In addition, we show that the average recall value is improved by 13.72% over CNN by using the Thermal Face-CNN proposed in this paper for the thermal image dataset. It is also shown that we found that Thermal Face-CNN performs better than CNN, MLP, and C-SVM when the precision is slightly more crucial than recall through F-measure.

2. Background and Related Work

Face detection is a field involving the detection of a face in an image. Algorithms for face detection judge whether or not the object in the picture is the face [5]. However, face liveness detection is a field in which the face presented is judged to be the real face or the fake face or no face. Therefore, face detection is a very different field from face liveness detection. For this reason, a paper related to face detection could not be compared with a paper related to face liveness detection. In the field of face liveness detection, there are three ways to imitate a real face: using a picture with that face, replaying a video with that face, and using a 3D face mask [6]. The method using the picture with the face involves printing the face on paper or displaying the face on a display. In order to solve this problem, studies have been carried out to explore ways to detect the real face using a photo-based dataset [6–9]. In addition, there have been studies into the use of video-based datasets to distinguish the real face from the fake face [7,10]. Further studies into ways to distinguish between the real face and the 3D face mask have also been conducted [11,12].

Many datasets can be used for face liveness detection: NUAA [8], ZJU Eyeblink [13], Idiap Print-attack [14], Idiap Replay-attack [10], CASIA FASD [15], MSU-MFSD [16], MSU RAFS [17], UVAD [18,19], MSU USSA [6], and so on. However, these datasets include data composed of RGB images. There are not enough datasets composed of thermal images. Therefore, research on face liveness detection with thermal images has been insufficient to date. Thermal images have already been used in research for face detection and pedestrian detection [20–23]. Thermal images can be obtained through the distribution of infrared rays, even at night when there is no visible light. Because RGB images have the disadvantage of being affected by the intensity of visible light, while thermal images have the advantage of being usable in places where there is no visible light, thermal images have been successfully applied in various fields. Therefore, it is necessary to compare the RGB image and the thermal image with regard to how much performance improvement is offered by the use of the thermal image in face liveness detection. For comparison, using an existing dataset would be ideal, but none of these contain information about temperature. Thus, a new dataset is needed.

Face liveness detection involves detecting the real face by analyzing the information obtained from the image. Therefore, previous studies on face liveness detection have been carried out using image processing methods. The support vector machine (SVM) is a classification algorithm that has been used to distinguish between the real and fake faces in face liveness detection [7,11]. As shown in these studies, SVM performs well in the area of classification. Of the SVM algorithms, the linear SVM finds the linear hyperplane with the largest margin [24]. The linear SVM assumes that classification can be performed by a line. However, there are cases where the data to be classified cannot be simply classified as a line. In order to solve this problem, research was carried out on nonlinear SVM using kernel functions [24]. The classification was proceeded using SVM on the abstraction information combining static features and dynamic features for face liveness detection in [7]. In addition, SVM learned the multispectral reflectance distribution information that can distinguish real human skin from images or objects meant to look like skin for face liveness detection in [11]. Previously, SVM used in face liveness detection learned to perfectly classify training data without error. However, there is another way to find a soft margin hyperplane that has the largest margins while allowing exceptional misclassification of the small amount of data in the learning data [3]. By using a soft margin hyperplane, we can find a hyperplane that is more generalizable without having an overfitting hyperplane on the learning data. Therefore, C-SVM, which is a nonlinear SVM using a soft margin hyperplane and more generalizable than the SVMs used in previous studies, was used in Section 4 to evaluate the performance of algorithms on the thermal image dataset.

The artificial neural network imitates human neurons [1]. In particular, MLP is one of the artificial neural networks used in image processing [25]. Image processing can be done through MLP, in which the information of pixels is inserted into the input layer, and the output layer outputs 0 and 1 with one node for binary classification. CNN [2], which is designed for effective image processing, is an algorithm that modifies MLP in a way that reduces weights and shares weights. There are studies that have effectively performed face liveness detection using CNN on the RGB image [7,26,27]. In addition, it is known that CNN is a more powerful algorithm for face liveness detection on the RGB image than SVM [26]. Furthermore, CNN can achieve 98.99% accuracy on the relatively easy RGB image dataset called NUAA [8], which means that CNN is superior to previous methods [26] and is state-of-the-art. An accuracy of 98.99% does not mean that this field is entirely conquered. There is a need to study more difficult face liveness detection by allowing multiple objects to be included simultaneously in an image and increasing a lot of computation with more pixels in an image. The thermal image can be used to do this because there have also been studies showing that CNN has been successfully used on the thermal image [20–22]. For these reasons, and because there is a need to properly process the thermal image used for face liveness detection with CNN, we used this algorithm in Section 4. Nevertheless, it is necessary to investigate an algorithm superior to CNN for face liveness detection based on the thermal image. The CNN algorithm and Thermal Face-CNN for face liveness detection are concretely described in Section 3 of this paper.

In addition to the support vector machine and the artificial neural network, the algorithms used for face liveness detection are diverse. A logistic regression model [8,28] was used to classify the real face and the fake face. In addition, as methods to identify the features of the image, local binary pattern [9,29] and Lambertian model [8] were used for face liveness detection. The local binary pattern is a method of extracting the feature of the image considering the difference of value relative to neighboring pixels on the basis of a pixel. By this method, the feature vector representing the feature of the image was extracted for face liveness detection [9]. Similarly, the Lambertian model is a method that has been studied for extracting information about the difference between the real face and fake face. Therefore, we can know that there has been a lot of research on how to extract image feature information in the related studies.

3. The Proposed Method

The proposed Thermal Face-CNN is an algorithm for face liveness detection based on CNN. In this algorithm, external knowledge for face liveness detection is inserted first, followed by CNN. In the proposed method, the artificial neural network part is the same as the existing CNN. CNN combines the convolutional layer, the pooling layer, and the fully connected layer. The number of convolutional layers, pooling layers, and fully connected layers vary depending on the number and type of pixels in the image. For visual convenience, an example of Thermal Face-CNN with two convolutional layers, two pooling layers, and one hidden layer is shown in Figure 1. The numbers of layers used are explained in Section 4.

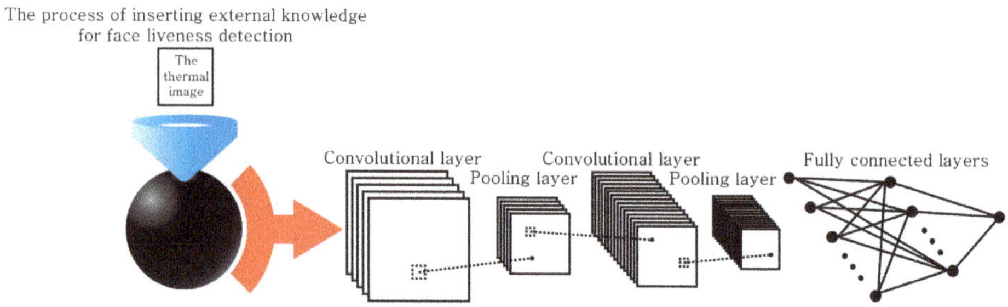

The process of inserting external knowledge for face liveness detection

The thermal image

Convolutional layer Pooling layer Convolutional layer Pooling layer Fully connected layers

Figure 1. Thermal face-convolutional neural network (Thermal Face-CNN).

First, knowledge is inserted for face liveness detection. After that, the data with external knowledge is calculated in the convolutional layer and transferred to the pooling layer. This can be repeated several times in order to process the complex image. Next, CNN passes the previously obtained information to the fully connected layer. Finally, CNN classifies the image in the output layer. The process of inserting external knowledge, the convolutional layer, the pooling layer, and fully connected layer are explained as the paper continues. The process of inserting external knowledge for face liveness detection can be accomplished by the process of inserting knowledge about the temperature that a human face can have. This can be represented as Equation (1).

$$h = \begin{cases} \textit{knowledge value} \times g & \textit{if down limit} \leq g \leq \textit{up limit} \\ g & \textit{Otherwise} \end{cases} \quad (1)$$

In Equation (1), g is the measured temperature value, and h is the input value to CNN. Equation (1) is a formula that multiplies the value between *up limit* and *down limit* by *knowledge value* so as to make use of the physiological knowledge of the mean body temperature of a person, which is between 36 and 37 degrees [30]. A pixel measuring a part of a real face must have a temperature value in this vicinity. The fact that there is a high probability that a pixel with a value close to 36 or 37 degrees in a measured thermal image is likely to represent a part of a real face can only be obtained from external knowledge, not from the data. In order to insert this knowledge into the artificial neural network, we make a remarkably different value than the measured value using Equation (1). In this case, the artificial neural network recognizes the temperature of this pixel as very different from the temperature measured at other pixels. If the *knowledge value* is 10, it is about ten times larger than the values of other pixels. Figure 2 shows an example of selecting 34 and 39 values near the human body temperature of 36 and 37 degrees, taking into account the errors that may occur during measurement. In Section 4, we conducted experiments setting various values of *knowledge value*, *up limit*, and *down limit*.

In the graph shown in the upper left of Figure 2, the vertical axis represents the temperature values. In the graph shown in the upper right of Figure 2, the external knowledge about the possibility that a part of an object measured by each pixel is a part of a real face and the possibility that it is not is expressed. Note that there are no quantitative values in the vertical axis shown in the upper right graph in Figure 2. All of the graphs of the horizontal axes shown in Figure 2 represent the pixel index. In the upper left graph in Figure 2, pixels 2 and 3 are data with different meanings from the graph on the upper right, but there is almost no quantitative difference. In order to emphasize this content, input data must be re-expressed so that there are distinct differences between the two different data: one might measure a part of a real face, and the other might not. To do so, *knowledge value* in Equation (1) is used. As shown in the graph in Figure 2, below, information is forced to be distributed in a specific region through a considerable difference between real values, and thermal information about the temperature value of the pixels measured is also expressed showing a minute difference. The differences in measured temperatures can be seen by comparing pixel 1 to pixel 3 and pixel 2 to pixel 4. The optimal knowledge value can be empirically found through experimentation.

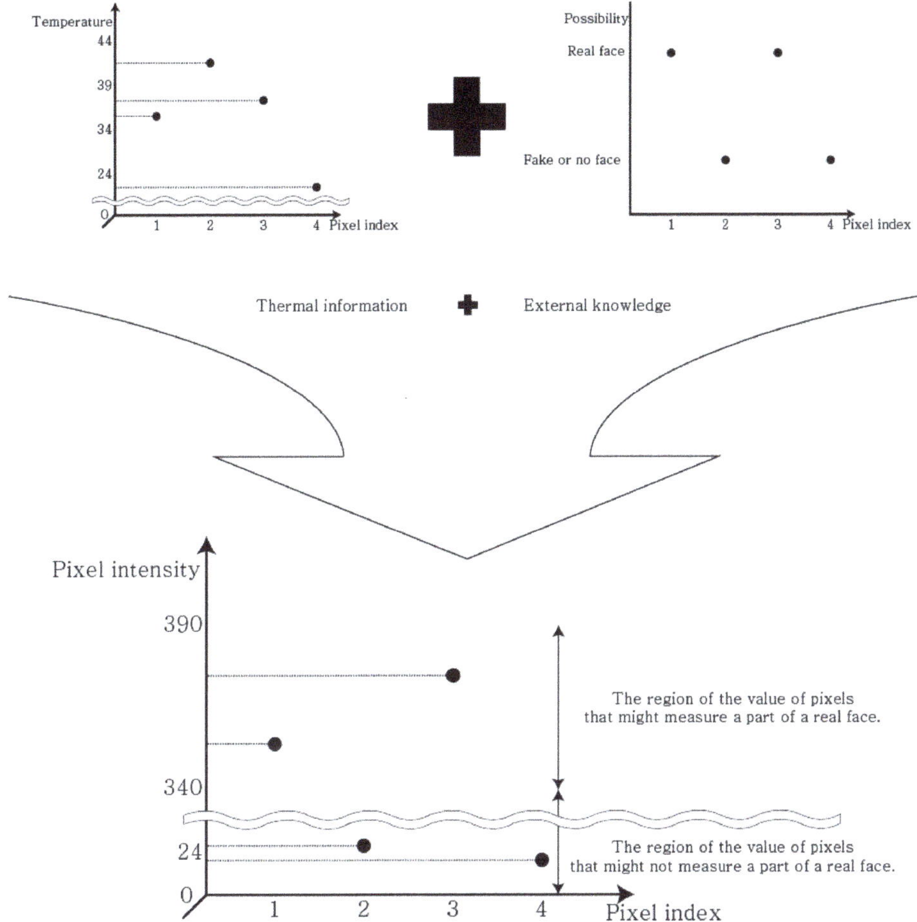

Figure 2. Example of the process of inserting external knowledge.

The convolutional layer serves to extract the complex features of the two-dimensional image [31]. The parameters of the convolutional layer are *kernel_size*, *filters*, and *stride*. *kernel_size* indicates the width and height of a kernel composed of learnable weights. *filters* represent the number of kernels, and *stride* is a parameter for extracting the characteristics of an image based on a certain interval. From the convolutional layer, we can extract the spatial information while sharing the weights [2]. Formal equations related to the convolutional layer are presented in [31]. The information calculated in the convolutional layer is transferred to the pooling layer.

Among the layers that make up CNN, the pooling layer induces spatial invariance by reducing the size of the feature map [32]. The parameters of the pooling layer are *pooling_size* and *stride*. *pooling_size* represents the size of the zone to be examined, such as *kernel_size*, a parameter of the convolutional layer discussed above. *stride* in the pooling layer serves the same purpose as the *stride* parameter of the convolutional layer. The max pooling layer has a function to find the maximum value in each region and to transfer it to the next layer [32]. Finally, the information is transferred to the fully connected layer through the convolutional layer and the pooling layer.

The fully connected layer is a type of layer used in MLP consisting of nodes completely connected to the nodes in each of the previous and subsequent layers [1].

4. Experiments

4.1. Data Collection and Experimental Environment Construction

The Flir C3 was used as the camera for collecting data. The camera has two lenses on the front: an RGB lens to obtain RGB images of 640 × 480 pixels and an infrared lens to obtain thermal images of 80 × 60 pixels. The information on the Flir C3 can be found at a website listed in Supplementary Materials at the end of this paper. We collected one RGB image and one thermal image in each scene to find suitable data for face liveness detection. Since a thermal image is better than an RGB image at night, we took images in indoor residential environments with visible light for accurate performance comparison. There were no conditions for the distance of the object. The faces in the dataset were used with and without a variety of accessories, such as glasses. No matter what, the face is covered by any object, which can cover anything except the eyes, nose, and mouth. We used the function of the Flir C3 that allows for the simultaneous operation of the two lenses. A total of 844 scenes were taken. The actual data used were 844 Excel files with temperature information collected from infrared lens and 2532 Excel files with R, G, and B information collected from RGB lens. In Figure 3, the images in the top row are RGB images, while the images in the bottom row are thermal images.

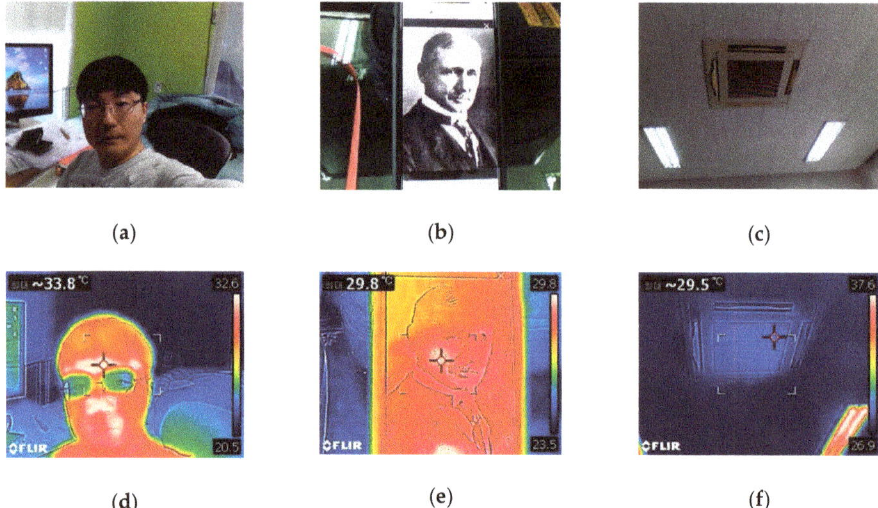

Figure 3. Data examples: (**a**) a real face taken by RGB lens; (**b**) a face on a display taken by RGB lens; (**c**) a ceiling air conditioner taken by RGB lens; (**d**) a real face taken by infrared lens; (**e**) a face on a display taken by infrared lens; (**f**) a ceiling air conditioner taken by infrared lens.

Figure 3a,d are RGB and thermal images with a real face present, respectively. Figure 3b,e are RGB and thermal images with a face on a display, respectively. Figure 3c,f shows images taken of a ceiling air conditioner with no face. In the thermal images, the color is obtained by the software in the thermal camera itself so that the measured temperature can be intuitively grasped visually. In Figure 3a,b,d,e, it can be seen that the outline of the heat distribution and the heat on the face from the display differ from those of the real face. The RGB face liveness detection dataset jongwoo (RFLDDJ) we created and the thermal face liveness detection dataset jongwoo (TFLDDJ) we created are available on the internet. In NUAA [8], the whole picture is completely filled with faces. However, in the RGB dataset we created, people and objects were shot in indoor living environments in order to increase the level of difficulty. In other words, multiple objects coexist in a single image in the datasets we made. The data are more difficult because a more general situation is assumed. The information of the datasets can be found at websites listed in the Supplementary Materials at the end of this paper.

The numbers of pixels differ between the two lenses. The RGB lens has 640 pixels horizontally and 480 pixels vertically, for a total of 307,200 pixels on an image. By contrast, the infrared lens has 80 pixels horizontally and 60 pixels vertically, for a total of 4800 pixels on an image. The numbers of pixels in images obtained by the two lenses differ by 64 times. However, the range of actually measured scenes is not much different. Figure 4 shows its example.

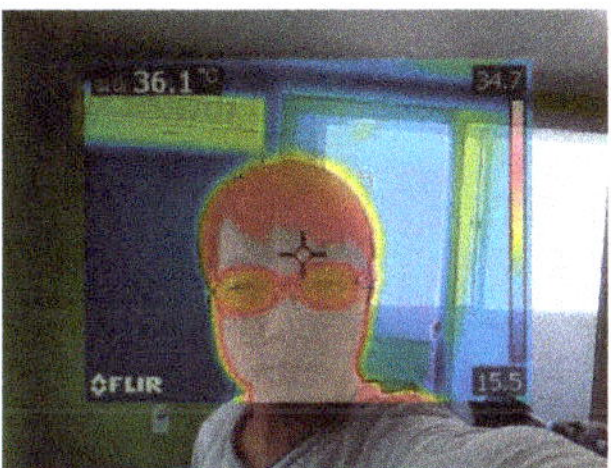

Figure 4. Comparison of the ranges of lenses.

As shown in Figure 4, the number of pixels has a difference of 64 times, but there is not much difference in the area to be taken. In addition, because the RGB lens and the infrared lens have different pixel sizes, and because there is a slight difference in the position of each lens on the camera, it is not clear how many pixels from the horizontal, vertical, top, and bottom sides should be cut for the same range of the scene. Therefore, it is impossible to capture the same extent of the range of the scene. For the correct experiment, if the real face is in a scene that the infrared lens cannot capture as an image, this image was removed from the experiment.

We use Adam [33], Dropout [34], and ReLu [35] to improve learning abilities when learning CNN and Thermal Face-CNN. The Adam algorithm reduces error by learning the weights existing in the artificial neural network. It is easier to execute than the back-propagation algorithm [36]. It is also more efficient and requires less memory [33]. Dropout prevents overfitting by allowing each node not to participate in the calculation randomly during the learning process [34]. Sigmoid [37] was used as an activation function in the output layer of all artificial neural networks used in the experiments except for C-SVM, and ReLu was used as an activation function of the hidden layer. As the pooling layer, the max pooling layer [32] is used. In addition, the probability of dropping each node is 10%. An intel core i7-7820X CPU was used as the hardware in the experiment, and the memory was DDR4 32G. The experiment was carried out using the Tensorflow [38] library, which has artificial neural network code. In the case of C-SVM, the sklearn.svm.svc library was used to carry out the experiment. The information of the library can be found at a website listed in the Supplementary Materials at the end of this paper.

Accuracy [4], recall [4], and precision [4] were mainly used as evaluation indices in the experiment. In this study, accuracy refers to how the actual value and predicted value are matched, regardless of the presence or absence of a real face. Recall is an index of how many images having the real face are judged to have the real face. Precision is also an index of how many images have the real face among those predicted to have the real face.

4.2. The Comparison of Face Liveness Detection between the RGB Image and Thermal Image

Before examining the performance of the proposed Thermal Face-CNN, we obtained accuracy, recall, and precision for each RGB image and thermal image dataset in order to identify the appropriate dataset for face liveness detection. For the comparison, we used CNN, MLP, and C-SVM. The left side of Table 1 shows the parameters of CNN applied to the RGB image dataset, and the right side of

Table 1 shows the parameters of CNN applied to the thermal image dataset. We empirically sought the values of the parameters that would make the error of the artificial neural network converge to zero.

Table 1. Convolutional neural network (CNN) parameters used in the RGB image dataset and the thermal image dataset.

Parameter	Kernel_ Size	Filters	Pool_ Size	Stride/ Nodes	Parameter	Kernel_ Size	Filters	Pool_ Size	Stride/ Nodes
1st con_	(15, 15)	150	N/A	(3, 3)	1st con_	(20, 20)	50	N/A	(3, 3)
1st pool_	N/A	N/A	(5, 5)	(1, 1)	1st pool_	N/A	N/A	(3, 3)	(2, 2)
2nd con_	(15, 15)	130	N/A	(3, 3)	2nd con_	(5, 5)	30	N/A	(1, 1)
2nd pool_	N/A	N/A	(5, 5)	(1, 1)	2nd pool_	N/A	N/A	(2, 2)	(1, 1)
3rd con_	(15, 15)	100	N/A	(2, 2)	input_	N/A	N/A	N/A	1920
3rd pool_	N/A	N/A	(3, 3)	(1, 1)	hidden_	N/A	N/A	N/A	120
4th con_	(5, 5)	80	N/A	(2, 2)	output_	N/A	N/A	N/A	1
4th pool_	N/A	N/A	(2, 2)	(1, 1)	N/A	N/A	N/A	N/A	N/A
input_	N/A	N/A	N/A	1920	N/A	N/A	N/A	N/A	N/A
1st hidden_	N/A	N/A	N/A	1536	N/A	N/A	N/A	N/A	N/A
2nd hidden_	N/A	N/A	N/A	1200	N/A	N/A	N/A	N/A	N/A
3rd hidden_	N/A	N/A	N/A	1000	N/A	N/A	N/A	N/A	N/A
output_	N/A	N/A	N/A	1	N/A	N/A	N/A	N/A	N/A

In Table 1, *nodes* refers to the number of nodes in the corresponding layer. Further, con_ means convolutional layer and pool_ means pooling layer. input_, hidden_, and output_ mean input layer, hidden layer, and output layer, respectively. The rest of the parameters are the same as those described in Section 3. In Table 1, the values in parentheses represent two values for the width and length of the kernel and pooling sequentially.

The parameter values for C-SVM used in the thermal image dataset are shown in Table 2.

Table 2. C-support vector machine (C-SVM) parameters used in the thermal image dataset.

Parameter	Error Penalty	Kernel	Gamma	Tolerance	Degree
Value	c	RBF or POLY	$1/n$_Features	0.001	3

In Table 2, c is an *error penalty* parameter, and we changed c when we experimented. RBF [39] or polynomial (POLY) [39] is used as *kernel*. *gamma* is the coefficient of *kernel*. In addition, *n_features* means the number of features and *tolerance* means stopping criterion. *degree* means the degree of the polynomial kernel function.

The parameters of the MLP used to learn the thermal images are shown in Table 3.

Table 3. Multi-layer neural network (MLP) parameters in the thermal image dataset.

Parameter	Input_	1st Hidden_	2nd Hidden_	3rd Hidden_	4th Hidden_	Output_
Nodes	4800	3000	2000	1500	1000	1

A total of 599 images in the RGB image dataset and thermal image dataset from image 1 to image 599 were used as training data, and the remaining 245 images were used for test data. There are 338 images of 844 images with the real face, and 506 images without the real face. In the training set are 225 images with the real face, and 113 images with the real face are in test set. In the training set were 374 images without the real face, and 132 images without the real face are in the test set. Table 4 shows the experimental results of CNN in the RGB image dataset and the thermal image dataset. Tables 5 and 6 show the experimental results of MLP and C-SVM in the thermal image dataset. The figures in the following tables, including Tables 4–6, were rounded to the fourth decimal place. Figures expressed as percentages in the following tables were rounded to the second decimal place.

Table 4. CNN's performance in the RGB image dataset and the thermal image dataset.

Index	In the RGB Image Dataset			Index	In the Thermal Image Dataset		
	Accuracy	Recall	Precision		Accuracy	Recall	Precision
Average	0.658	0.4779	0.6871	Average	0.7816	0.6996	0.8022
The best	0.6898	0.5752	0.7342	The best	0.8367	0.7876	0.8476

Table 5. MLP's performance in the thermal image dataset.

Index	MLP		
	Accuracy	Recall	Precision
Average	0.7551	0.4991	0.9431
The best	0.7837	0.5664	0.9524

Table 6. C-SVM's performance in the thermal image dataset.

kernel	c	Accuracy	Recall	Precision	Kernel	c	Accuracy	Recall	Precision
RBF	0.7	0.5429	0.0088	1	POLY	0.06	0.7388	0.6195	0.7692
	0.8	0.8163	0.9381	0.7361		0.07	0.7388	0.6195	0.7692
	0.81	0.8082	0.9381	0.726		0.07	0.7388	0.6195	0.7692
	0.82	0.8204	0.9646	0.7315		0.08	0.7388	0.6195	0.7692
	0.83	0.8204	0.9646	0.7315		0.08	0.7388	0.6195	0.7692
	0.84	0.8122	0.9646	0.7219		0.09	0.7388	0.6195	0.7692
	0.85	0.8082	0.9646	0.7171		0.1	0.7388	0.6195	0.7692
	0.86	0.8082	0.9646	0.7171		0.11	0.7388	0.6195	0.7692
	0.87	0.8082	0.9646	0.7171		0.13	0.7388	0.6195	0.7692
	0.88	0.8082	0.9646	0.7171		0.14	0.7388	0.6195	0.7692
	0.89	0.8122	0.9646	0.7219		0.17	0.7388	0.6195	0.7692
	0.9	0.8122	0.9646	0.7219		0.2	0.7388	0.6195	0.7692
	0.91	0.8163	0.9646	0.7267		0.25	0.7388	0.6195	0.7692
	0.92	0.8204	0.9646	0.7315		0.3	0.7388	0.6195	0.7692
	0.93	0.8204	0.9646	0.7315		0.33	0.7388	0.6195	0.7692
	0.94	0.8204	0.9646	0.7315		0.4	0.7388	0.6195	0.7692
	0.95	0.8122	0.9469	0.7279		0.5	0.7388	0.6195	0.7692
	0.96	0.8122	0.9381	0.731		0.6	0.7388	0.6195	0.7692
	0.97	0.8163	0.9381	0.7361		0.7	0.7388	0.6195	0.7692
	0.98	0.8163	0.9381	0.7361		0.8	0.7388	0.6195	0.7692
	0.99	0.8204	0.9381	0.7413		0.9	0.7388	0.6195	0.7692
	1	0.8245	0.9381	0.7465		1	0.7388	0.6195	0.7692
	1.5	0.8204	0.9292	0.7447		1.5	0.7388	0.6195	0.7692
	2	0.8204	0.9292	0.7447		2	0.7388	0.6195	0.7692
	2.5	0.8204	0.9292	0.7447		2.5	0.7388	0.6195	0.7692

In Tables 4 and 5, "The best" refers to the highest values. "Average" means the average value. In order to obtain the information shown in Table 4, five CNNs in the RGB image dataset and 20 CNNs in the thermal image dataset were implemented with the same parameters. Because the combinations of weights obtained when the neural network is learned with the same parameters are always different and show different performances, we repeated the experiment 20 times in order to obtain the average performance of the general accuracy, recall, and precision values. However, in the RGB image dataset, the number of pixels contained in each image was 907,200, which required a substantial amount of computation. Therefore, 20 CNNs were learned in the thermal image dataset, but only five CNNs were learned in the RGB image dataset. To obtain Table 5, five MLPs were learned because MLP requires a large amount of computation. To evaluate C-SVM's performance in Table 6, we obtained one C-SVM on each parameter setting. The values of accuracy, recall, and precision shown in Table 4, which were obtained using the thermal image dataset, are higher than those of the RGB image dataset. It can be seen from the above that, on CNN, the thermal image is more suitable than the RGB image.

In the case of MLP, since there is 907,200-pixel information per RGB image, the number of nodes in the input layer should also be 907,200. We tried to implement an MLP with about 900,000 nodes in the input layer, but the hardware limitations made it impossible to calculate. Further, the C-SVM was learned using the parameters shown in Table 2, but it was determined that there was no real face for all the test data, because it was not learned properly. However, as shown in Tables 5 and 6, MLP and C-SVM can be learned because of the small number of pixels in a thermal image data. Through comparing Tables 4–6, it can be seen that good performance can be obtained by the thermal image data.

4.3. Performance Comparison of CNN, C-SVM, and Thermal Face-CNN

Section 4.2 showed that the thermal image is better than the RGB image. In Section 4.3, we applied the Thermal Face-CNN proposed in this paper to the thermal image with superior performance for face liveness detection than the RGB image, and we compared its performance with those of the other algorithms. We used the same parameters of CNN on the thermal image dataset for Thermal Face-CNN. We also constructed 20 Thermal Face-CNNs with the same parameter setting as used in the experiment on 20 CNNs, shown in Table 4. The accuracy, recall, and precision values of Thermal Face-CNNs are shown in Tables 7–12. Parenthetical values in these tables indicate *knowledge value*, *up limit*, and *down limit* values, sequentially.

Table 7. Thermal Face-CNN accuracy, recall, and precision values 1.

Index	(10, 39, 34)			Index	(10, 40, 34)		
	Accuracy	**Recall**	**Precision**		**Accuracy**	**Recall**	**Precision**
Average	0.7967	0.7726	0.784	Average	0.7957	0.7602	0.7901
The best	0.8245	0.8584	0.8173	The best	0.8204	0.8407	0.8235
A_im (%)	1.93	10.44	−2.27	A_im (%)	1.77	7.97	−1.53
M_im (%)	−1.46	8.99	−3.58	M_im (%)	−1.99	6.32	−2.93

Table 8. Thermal Face-CNN accuracy, recall, and precision values 2.

Index	(10, 41, 34)			Index	(10, 39, 35)		
	Accuracy	**Recall**	**Precision**		**Accuracy**	**Recall**	**Precision**
Average	0.7894	0.7491	0.7851	Average	0.7929	0.7385	0.7986
The best	0.8286	0.8142	0.8286	The best	0.8327	0.8142	0.8391
A_im (%)	0.99	6.61	−2.18	A_im (%)	1.43	5.27	−0.45
M_im (%)	−0.98	3.27	−2.29	M_im (%)	−0.48	3.27	−1.01

Table 9. Thermal Face-CNN accuracy, recall, and precision values 3.

Index	(10, 39, 33)			Index	(100, 39, 34)		
	Accuracy	**Recall**	**Precision**		**Accuracy**	**Recall**	**Precision**
Average	0.7845	0.7863	0.758	Average	0.7843	0.7535	0.7731
The best	0.8245	0.8938	0.8218	The best	0.8327	0.8319	0.8103
A_im (%)	0.37	12.39	−5.51	A_im (%)	0.34	7.15	−3.76
M_im (%)	−1.46	13.48	−3.05	M_im (%)	−0.48	5.33	−4.6

Table 10. Thermal Face-CNN accuracy, recall, and precision values 4.

Index	(−5, 39, 34)			Index	(−10, 39, 34)		
	Accuracy	**Recall**	**Precision**		**Accuracy**	**Recall**	**Precision**
Average	0.8151	0.7956	0.8027	Average	0.8033	0.7903	0.7853
The best	0.8367	0.8407	0.8515	The best	0.8367	0.8673	0.8214
A_im (%)	4.29	13.72	0.06	A_im (%)	2.7	11.47	−2.16
M_im (%)	0	6.74	0.46	M_im (%)	0	9.18	−3.19

Table 11. Thermal Face-CNN accuracy, recall, and precision values 5.

Index	(−100, 39, 34)			Index	(5, 39, 34)		
	Accuracy	Recall	Precision		Accuracy	Recall	Precision
Average	0.7912	0.7726	0.7755	Average	0.7939	0.7429	0.7972
The best	0.8163	0.8761	0.8367	The best	0.8367	0.8496	0.8349
A_im (%)	1.23	10.43	−3.33	A_im (%)	1.57	6.19	−0.63
M_im (%)	−2.43	11.24	−1.28	M_im (%)	0	7.87	−1.5

Table 12. Thermal Face-CNN accuracy, recall, and precision values 6.

Index	(1000, 39, 34)		
	Accuracy	Recall	Precision
Average	0.7294	0.6372	0.7399
The best	0.7918	0.7434	0.8298
A_im (%)	−7.16	−9.79	−8.42
M_im (%)	−5.67	−5.95	−2.15

"The best" and "Average" in Tables 7–12 mean the highest value and average value, respectively. In Tables 7–12, A_im (%) means how much the average value is improved in comparison with CNN, and M_im (%) means how much the maximum value is improved in comparison with CNN. For example, A_im (%) and M_im (%) are obtained by average and the best values in the right side of Tables 4 and 7, Tables 8–12. The information on all the experimental results can be found at websites listed in the Supplementary Materials found at the end of this paper.

When the *knowledge value* is 10 in the Thermal Face-CNNs described in Tables 7 and 8 and the left side of Table 9, the values of accuracy, recall, and precision are obtained as changes occur to the values of the *up limit* and *down limit*. When the *up limit* and *down limit* are 39 and 33, respectively, the average recall value has the greatest increase, by 12.39%. When the *up limit* and *down limit* values are 39 and 34, respectively, the average recall value is increased by 10.44%. When the *up limit* and *down limit* are 40 and 34, respectively, the average recall value is increased by 7.97%, and the average precision value is decreased slightly by −1.53%. In addition, when the *up limit* and *down limit* are 41 and 34, respectively, the average recall is increased by 6.61%, and the precision is decreased by −2.18%. When the values of the *up limit* and *down limit* are 39 and 35, respectively, the amount of the increment of recall is reduced the best.

The Thermal Face-CNNs described on the left side of Table 7 and the right side of Tables 9 and 10, Tables 11 and 12 show the amount by which the performance changed when the *up limit* and *down limit* are 39 and 34, respectively, and when the *knowledge value* is changed. Table 12 shows that much lower performance can be achieved with Thermal Face-CNN than with CNN. The Thermal Face-CNN used to obtain the data in Table 12 has the same parameters as the Thermal Face-CNNs used to obtain the data in the left side of Table 7, except for the fact that the *knowledge value* is 1,000. Therefore, a huge *knowledge value* shows that performance can be rather reduced. The best performance was obtained by increasing the average recall value by 13.72% when the *knowledge value* was −5, and the second-best average recall value was increased by 11.47% when the *knowledge value* was −10. In addition, when the *knowledge value* was 10, the third-best performance was obtained by increasing the average recall value by 10.44%. When the *knowledge value* was −100, the average recall value was increased by 10.43%, which was the fourth-best performance.

Except for Table 12, the average recall values of the Thermal Face-CNN having external knowledge about the temperature of the real face in Tables 7–11 show that the average recall value and the best recall value are better than the CNN shown in the right side of Table 4. An increase of the recall value means that the Thermal Face-CNN has detected more data having the real face than CNN. It can be seen that CNN and Thermal Face-CNN are not significantly different in terms of accuracy and precision when we compare the values in the right sides of Tables 4 and 7, Tables 8–11. Looking at the performance of Thermal Face-CNN that obtained the best performance, in the left side of Table 10, we can see that Thermal Face-CNN was not reduced at all. Therefore, Thermal Face-CNN is superior to CNN in all indices.

The performance obtained by Thermal Face-CNN must be compared with the accuracy, recall, and precision values recorded in Tables 5 and 6 quantitatively. Table 10 shows that the method with the highest accuracy is 0.8367 on Thermal Face-CNN. In addition, the results in Table 6 show that C-SVM is the method with the highest recall. Further, Table 5 shows that MLP is the method with the highest precision. However, MLP is a relatively bad way to detect the real face because the recall value is too small. Thermal Face-CNN has the best accuracy and more balance between recall and precision than MLP and C-SVM. For accurate performance evaluation, F-measure [40] is used. F-measure is a widely used index that quantitatively evaluates performance by simultaneously considering recall and precision. F-measure is shown in Equation (2).

$$F\text{-}measure = \frac{(\beta^2 + 1) \times precision \times recall}{\beta^2 \times precision + recall} \tag{2}$$

β is a positive real number or zero. Also *precision*, *recall*, and *F-measure* are the values of precision, recall, and F-measure, respectively. A larger F_measure value means a better algorithm. When β is one, the most frequently used F-measure formula appears in Equation (3).

$$F\text{-}measure\_1 = \frac{2 \times precision \times recall}{precision + recall} \tag{3}$$

F-measure_1 in Equation (3) means the value of F-measure when β is one. As shown in Equation (4), *difference* denotes the difference value of F-measures of the Thermal Face-CNN and C-SVM; Thermal Face-CNN obtained 0.8327 accuracy, 0.8407 recall, 0.8051 precision, and C-SVM obtained 0.8245 accuracy, 0.9381 recall, 0.7465 precision corresponding to Table 6.

$$difference = \frac{(\beta^2 + 1) \times 0.8051 \times 0.8407}{\beta^2 \times 0.8051 + 0.8407} - \frac{(\beta^2 + 1) \times 0.7465 \times 0.9381}{\beta^2 \times 0.7465 + 0.9381} \tag{4}$$

When the *difference* is zero, the β value is 0.8885, meaning that the two *f-measure* values are the same. When β is greater than or equal to 0 and less than 0.8885, then Thermal Face-CNN is better. By contrast, when β is greater than 0.8885, C-SVM is better. You can find the corresponding conditions by obtaining equations in the same way for several Thermal Face-CNNs. It is trivial to find β that makes *difference* zero when the parameters are different. Nevertheless, it is important to show that the Thermal Face-CNN is superior by listing the F-measures obtained at commonly used β values of 0.5 and 2. Table 13 shows it.

In Table 13, "Average F-measure" means the F-measure using average recall and average precision in the left side of Table 10. When β is 2, F-measure means that F-measure weighs recall higher than precision. When β is 0.5, F-measure means that F-measure weighs recall lower than precision. Therefore, we can see that Thermal Face-CNN is best when precision has more weight than recall. Precision is more important than recall when the reliability of the algorithm is important. Therefore, Thermal Face-CNN is good for this situation.

In addition to the comparison based on accuracy, recall, precision, and F-measure, it is shown that the CNN-based proposed algorithm is superior to CNN and has similar performance with the others on receiver operating characteristic (ROC) graph [41] in Figure 5. Parenthetical values in Figure 5 indicate *knowledge value*, *up limit*, and *down limit* values, sequentially.

Table 13. C-SVM's and Thermal Face-CNN's F-measure comparison.

			Which algorithm is superior?	
c	F-measure on C-SVM with RBF and $\beta = 0.5$	F-measure on C-SVM with RBF and $\beta = 2$	When Thermal Face-CNN has (−5, 39, 34) and $\beta = 0.5$, Average F-measure = 0.8013	When Thermal Face-CNN has (−5, 39, 34) and $\beta = 2$, Average F-measure = 0.797
0.7	0.0425	0.011	Thermal Face-CNN	Thermal Face-CNN
0.8	0.7692	0.8893	Thermal Face-CNN	C-SVM
0.81	0.7604	0.8863	Thermal Face-CNN	C-SVM
0.82	0.7686	0.9068	Thermal Face-CNN	C-SVM
0.83	0.7686	0.9068	Thermal Face-CNN	C-SVM
0.84	0.7602	0.9038	Thermal Face-CNN	C-SVM
0.85	0.7559	0.9023	Thermal Face-CNN	C-SVM
0.86	0.7559	0.9023	Thermal Face-CNN	C-SVM
0.87	0.7559	0.9023	Thermal Face-CNN	C-SVM
0.88	0.7559	0.9023	Thermal Face-CNN	C-SVM
0.89	0.7602	0.9038	Thermal Face-CNN	C-SVM
0.9	0.7602	0.9038	Thermal Face-CNN	C-SVM
0.91	0.7644	0.9053	Thermal Face-CNN	C-SVM
0.92	0.7686	0.9068	Thermal Face-CNN	C-SVM
0.93	0.7686	0.9068	Thermal Face-CNN	C-SVM
0.94	0.7686	0.9068	Thermal Face-CNN	C-SVM
0.95	0.7632	0.8932	Thermal Face-CNN	C-SVM
0.96	0.7648	0.8878	Thermal Face-CNN	C-SVM
0.97	0.7692	0.8893	Thermal Face-CNN	C-SVM
0.98	0.7692	0.8893	Thermal Face-CNN	C-SVM
0.99	0.7738	0.8908	Thermal Face-CNN	C-SVM
1	0.7783	0.8923	Thermal Face-CNN	C-SVM
1.5	0.7755	0.8853	Thermal Face-CNN	C-SVM

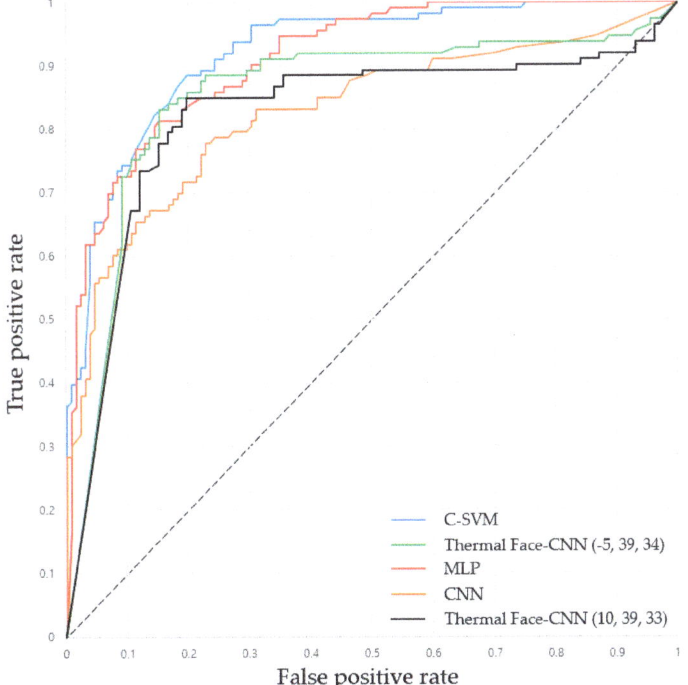

Figure 5. Receiver operating characteristic (ROC) graph.

'A' line is better than 'B' line if 'A' line is closer to the northwest than 'B' line in ROC graph. The blue line in Figure 5 shows the performance of C-SVM, the green and black lines show the performance of Thermal Face-CNN, the red line shows the performance of MLP, and the orange line shows the performance of CNN. To obtain Figure 5, we used the parameters having the best performance: MLP which has an accuracy of 0.7837, a recall of 0.5664, and a precision of 0.9412 and the CNN which has an accuracy of 0.8367, a recall of 0.7876, and a precision of 0.8476 and the best performance among a *up limit* value of 39, and a *down limit* value of 34 in Thermal Face-CNN which has an accuracy of 0.8327, a recall of 0.8407, a precision of 0.8051, a *knowledge value* value of−5, a *up limit* value of 39, and a *down limit* value of 34 and the best performance among a *knowledge value* of 10 in Thermal Face-CNN which has an accuracy of 0.8245, a recall of 0.8496, a precision of 0.7869, a *knowledge value* value of 10, a *up limit* value of 39, and a *down limit* value of 33 and C-SVM which has a *c* value of 1 are used. As shown in Figure 5, Thermal Face-CNN has the dramatic performance improvement compared to CNN, and the Thermal Face-CNN's performance is close to that of MLP and C-SVM. In this paper, we argue that Thermal Face-CNN is better when precision is more important than recall. However, ROC graph does not directly consider precision because it uses true positive rate and false positive rate, which are not precision. Nonetheless, the ROC graph shows that Thermal Face-CNN is superior to CNN.

5. Conclusions and Future Works

Face liveness detection is an important field that allows for information about a real person to be communicated when communicating security. In this paper, face liveness detection was performed in indoor residential environment using the fact that thermal patterns on a face in a display and a photograph differ from those on the real face. First, we quantitatively compared the performance of the thermal image with the RGB image. It has been shown that the thermal image is more suitable for face liveness detection because CNN has the best performance, with an accuracy of 0.6898, a recall of 0.5752, a precision of 0.7342 on the RGB image dataset, and an accuracy of 0.8367, a recall of 0.7876, and a precision of 0.8476 on the thermal image dataset. We also propose Thermal Face-CNN, which has external knowledge about the real face temperature in the existing CNN algorithm and compares it with CNN. The performance of the best-performing Thermal Face-CNN is equal to or better than CNN. Furthermore, we used the F-measure to identify the condition in which the Thermal Face-CNN performs better than the C-SVM.

Based on the results in this paper, we hope that Thermal Face-CNN with the thermal image is used to detect malicious tricks to imitate the face. This paper shows that it is possible to insert external knowledge by adjusting the value of a particular real number range. Therefore, it is expected that the application algorithms that have knowledge in various fields will emerge.

In this study, the experiment was conducted using 844 scenes. Nevertheless, as the number of data increases, it becomes more feasible to use face liveness detection in more general situations. Therefore, there is a need to collect thermal images in the future. Moreover, due to the difference between the RGB lens and the infrared lens, the images measured differ in terms of pixel size, the number of pixels, and the range of the scene. Therefore, there is a need to construct datasets with fewer differences between the RGB and thermal image. Because the experiments of all the possible combinations of the parameters in the algorithms were not done, the comparisons are not conclusive. Therefore, it is necessary to accurately identify the optimal parameters combination that obtains the highest accuracy, recall, precision, F-measure value through additional experimentation.

Supplementary Materials: The information about Flir C3 is available online at https://www.flir.com/products/c3/. The RGB image dataset is available online at https://www.researchgate.net/publication/328297217_RGB_Face_Liveness_Detection_Dataset_JongwooRFLDDJ. The thermal image dataset is available online at https://www.researchgate.net/publication/327716173_Thermal_Face_Liveness_Detection_Dataset_JongwooTFLDDJ. SVC is available online at https://scikit-learn.org/stable/modules/generated/sklearn.svm.SVC.html. All the experimental results are available online at https://www.researchgate.net/publication/330359019_symmetry_experiments.

The raw images are available online at https://www.researchgate.net/publication/330382261_Raw_images_for_Face_Liveness_Detection_Using_Thermal_Face-CNN_with_External_Knowledge.

Author Contributions: Conceptualization, J.S.; data curation, J.S.; formal analysis, J.S.; investigation, J.S.; methodology, J.S.; project administration, I.-J.C.; resources, J.S.; software, J.S.; supervision, I.-J.C.; validation, I.-J.C. and J.S.; writing—original draft, J.S.; writing—review & editing, I.-J.C. and J.S.

Funding: This research received no external funding.

Conflicts of Interest: The authors declare no conflict of interest.

References

1. Svozil, D.; Kvasnicka, V.; Pospichal, J. Introduction to multi-layer feed-forward neural networks. *Chemom. Intell. Lab. Syst.* **1997**, *39*, 43–62. [CrossRef]
2. Lawrence, S.; Giles, C.L.; Tsoi, A.C.; Back, A.D. Face recognition: A convolutional neural-network approach. *IEEE Trans. Neural Netw.* **1997**, *8*, 98–113. [CrossRef] [PubMed]
3. Cortes, C.; Vapnik, V. Support-vector networks. *Mach. Learn.* **1995**, *20*, 273–297. [CrossRef]
4. Powers, D.M. Evaluation: From Precision, Recall and F-Measure to ROC, Informedness, Markedness and Correlation. *J. Mach. Learn. Technol.* **2011**, *2*, 37–63.
5. Jiang, H.; Learned-Miller, E. Face detection with the faster R-CNN. In Proceedings of the 2017 12th IEEE International Conference on Automatic Face & Gesture Recognition (FG 2017), Washington, DC, USA, 30 May–3 June 2017; pp. 650–657.
6. Patel, K.; Han, H.; Jain, A.K. Secure face unlock: Spoof detection on smartphones. *IEEE Trans. Inf. Forensics Secur.* **2016**, *11*, 2268–2283. [CrossRef]
7. Wu, L.; Xu, Y.; Xu, X.; Qi, W.; Jian, M. A face liveness detection scheme to combining static and dynamic features. In Proceedings of the Chinese Conference on Biometric Recognition, Chengdu, China, 14–16 October 2016; Springer: Cham, Switzerland, 2016; pp. 628–636.
8. Tan, X.; Li, Y.; Liu, J.; Jiang, L. Face liveness detection from a single image with sparse low rank bilinear discriminative model. In Proceedings of the European Conference on Computer Vision, Crete, Greece, 5–11 September 2010; Springer: Berlin/Heidelberg, Germany, 2010; pp. 504–517.
9. Kim, G.; Eum, S.; Suhr, J.K.; Kim, D.I.; Park, K.R.; Kim, J. Face liveness detection based on texture and frequency analyses. In Proceedings of the 2012 5th IAPR International Conference on Biometrics (ICB), New Delhi, India, 29 March–1 April 2012; pp. 67–72.
10. Chingovska, I.; Anjos, A.; Marcel, S. On the effectiveness of local binary patterns in face anti-spoofing. In Proceedings of the International Conference of Biometrics Special Interest Group (BIOSIG), Darmstadt, Germany, 6–7 September 2012.
11. Zhang, Z.; Yi, D.; Lei, Z.; Li, S.Z. Face liveness detection by learning multispectral reflectance distributions. In Proceedings of the 2011 IEEE International Conference on Automatic Face & Gesture Recognition and Workshops (FG 2011), Santa Barbara, CA, USA, 21–25 March 2011; pp. 436–441.
12. Erdogmus, N.; Marcel, S. Spoofing face recognition with 3D masks. *IEEE Trans. Inf. Forensics Secur.* **2014**, *9*, 1084–1097. [CrossRef]
13. Pan, G.; Sun, L.; Wu, Z.; Lao, S. Eyeblink-Based Anti-Spoofing in Face Recognition from a Generic Webcamera. In Proceedings of the 2007 11th IEEE International Conference on Computer Vision(ICCV), Rio de Janeiro, Brazil, 14–20 October 2007; pp. 1–8.
14. Anjos, A.; Marcel, S. Counter-measures to photo attacks in face recognition: A public database and a baseline. In Proceedings of the 2011 International Joint Conference on Biometrics (IJCB), Washington, DC, USA, 11–13 October 2011; pp. 1–7.
15. Zhang, Z.; Yan, J.; Liu, S.; Lei, Z.; Yi, D.; Li, S.Z. A face antispoofing database with diverse attacks. In Proceedings of the 2012 5th IAPR International Conference on Biometrics (ICB), New Delhi, India, 29 March–1 April 2012; pp. 26–31.
16. Wen, D.; Han, H.; Jain, A.K. Face spoof detection with image distortion analysis. *IEEE Trans. Inf. Forensics Secur.* **2015**, *10*, 746–761. [CrossRef]
17. Patel, K.; Han, H.; Jain, A.K.; Ott, G. Live face video vs. spoof face video: Use of moiré patterns to detect replay video attacks. In Proceedings of the 2015 International Conference on Biometrics (ICB), Phuket, Thailand, 19–22 May 2015; pp. 98–105.

18. Pinto, A.; Schwartz, W.R.; Pedrini, H.; Rocha, A.D. Using visual rhythms for detecting video-based facial spoof attacks. *IEEE Trans. Inf. Forensics Secur.* **2015**, *10*, 1025–1038. [CrossRef]
19. Pinto, A.; Pedrini, H.; Schwartz, W.R.; Rocha, A. Face spoofing detection through visual codebooks of spectral temporal cubes. *IEEE Trans. Image Process.* **2015**, *24*, 4726–4740. [CrossRef] [PubMed]
20. König, D.; Adam, M.; Jarvers, C.; Layher, G.; Neumann, H.; Teutsch, M. Fully convolutional region proposal networks for multispectral person detection. In Proceedings of the 2017 IEEE Conference on Computer Vision and Pattern Recognition Workshops (CVPRW), Honolulu, HI, USA, 21–26 July 2017; pp. 243–250.
21. Zhang, X.; Chen, G.; Saruta, K.; Terata, Y. Deep Convolutional Neural Networks for All-Day Pedestrian Detection. In *Information Science and Applications*; Springer: Singapore, 2017; pp. 171–178.
22. Baek, J.; Hong, S.; Kim, J.; Kim, E. Efficient pedestrian detection at nighttime using a thermal camera. *Sensors* **2017**, *17*, 1850. [CrossRef] [PubMed]
23. Kwaśniewska, A.; Rumiński, J. Face detection in image sequences using a portable thermal camera. In Proceedings of the 13th Quantitative Infrared Thermography Conference, Quebec City, QC, Canada, 4–8 July 2016.
24. Burges, C.J. A tutorial on support vector machines for pattern recognition. *Data Min. Knowl. Discov.* **1998**, *2*, 121–167. [CrossRef]
25. Peña, J.; Gutiérrez, P.; Hervás-Martínez, C.; Six, J.; Plant, R.; López-Granados, F. Object-based image classification of summer crops with machine learning methods. *Remote Sens.* **2014**, *6*, 5019–5041. [CrossRef]
26. Alotaibi, A.; Mahmood, A. Deep face liveness detection based on nonlinear diffusion using convolution neural network. *Signal Image Video Process.* **2017**, *11*, 713–720. [CrossRef]
27. Akbulut, Y.; Şengür, A.; Budak, Ü.; Ekici, S. Deep learning based face liveness detection in videos. In Proceedings of the 2017 International Artificial Intelligence and Data Processing Symposium (IDAP), Malatya, Turkey, 16–17 September 2017; pp. 1–4.
28. Peixoto, B.; Michelassi, C.; Rocha, A. Face liveness detection under bad illumination conditions. In Proceedings of the 2011 18th IEEE International Conference on Image Processing (ICIP), Brussels, Belgium, 11–14 September 2011; pp. 3557–3560.
29. Boulkenafet, Z.; Komulainen, J.; Hadid, A. Face spoofing detection using colour texture analysis. *IEEE Trans. Inf. Forensics Secur.* **2016**, *11*, 1818–1830. [CrossRef]
30. Sund-Levander, M.; Forsberg, C.; Wahren, L.K. Normal oral, rectal, tympanic and axillary body temperature in adult men and women: A systematic literature review. *Scand. J. Caring Sci.* **2002**, *16*, 122–128. [CrossRef] [PubMed]
31. Cong, J.; Xiao, B. Minimizing computation in convolutional neural networks. In Proceedings of the International Conference on Artificial Neural Networks, Hamburg, Germany, 15–19 September 2014; Springer: Cham, Switzerland, 2014; pp. 281–290.
32. Scherer, D.; Müller, A.; Behnke, S. Evaluation of pooling operations in convolutional architectures for object recognition. In *Artificial Neural Networks—ICANN 2010*; Springer: Berlin/Heidelberg, Germany, 2010; pp. 92–101.
33. Kingma, D.P.; Ba, J. Adam: A method for stochastic optimization. *arXiv* **2014**, arXiv:1412.6980.
34. Srivastava, N.; Hinton, G.; Krizhevsky, A.; Sutskever, I.; Salakhutdinov, R. Dropout: A simple way to prevent neural networks from overfitting. *J. Mach. Learn. Res.* **2014**, *15*, 1929–1958.
35. LeCun, Y.; Bengio, Y.; Hinton, G. Deep learning. *Nature* **2015**, *521*, 436–444. [CrossRef] [PubMed]
36. Rumelhart, D.E.; Hinton, G.E.; Williams, R.J. Learning representations by back-propagating errors. *Nature* **1986**, *323*, 533–536. [CrossRef]
37. Jain, A.K.; Mao, J.; Mohiuddin, K.M. Artificial neural networks: A tutorial. *Computer* **1996**, *29*, 31–44. [CrossRef]
38. Abadi, M.; Barham, P.; Chen, J.; Chen, Z.; Davis, A.; Dean, J.; Devin, M.; Ghemawat, S.; Irving, G.; Isard, M.; et al. TensorFlow: A system for large-scale machine learning. In Proceedings of the 12th USENIX Conference on Operating Systems Design and Implementation, Savannah, GA, USA, 2–4 November 2016; Volume 16, pp. 265–283.
39. Chang, Y.-W.; Hsieh, C.-J.; Chang, K.-W.; Ringgaard, M.; Lin, C.-J. Training and testing low-degree polynomial data mappings via linear SVM. *J. Mach. Learn. Res.* **2010**, *11*, 1471–1490.

40. Sokolova, M.; Japkowicz, N.; Szpakowicz, S. Beyond accuracy, F-score and ROC: A family of discriminant measures for performance evaluation. In Proceedings of the Australasian Joint Conference on Artificial Intelligence, Hobart, Australia, 4–8 December 2006; Springer: Berlin/Heidelberg, Germany, 2006; pp. 1015–1021.
41. Fawcett, T. An introduction to ROC analysis. *Pattern Recognit. Lett.* **2006**, *27*, 861–874. [CrossRef]

Article

Emotion Classification Using a Tensorflow Generative Adversarial Network Implementation

Traian Caramihale, Dan Popescu * and Loretta Ichim

Department of Control Engineering and Industrial Informatics, University Politehnica of Bucharest, 060042 Bucharest, Romania; traian90@gmail.com (T.C.); loretta.ichim@upb.ro (L.I.)

* Correspondence: dan.popescu@upb.ro; Tel.: +40-76-621-8363

Received: 11 July 2018; Accepted: 17 September 2018; Published: 19 September 2018

Abstract: The detection of human emotions has applicability in various domains such as assisted living, health monitoring, domestic appliance control, crowd behavior tracking real time, and emotional security. The paper proposes a new system for emotion classification based on a generative adversarial network (GAN) classifier. The generative adversarial networks have been widely used for generating realistic images, but the classification capabilities have been vaguely exploited. One of the main advantages is that by using the generator, we can extend our testing dataset and add more variety to each of the seven emotion classes we try to identify. Thus, the novelty of our study consists in increasing the number of classes from N to 2N (in the learning phase) by considering real and fake emotions. Facial key points are obtained from real and generated facial images, and vectors connecting them with the facial center of gravity are used by the discriminator to classify the image as one of the 14 classes of interest (real and fake for seven emotions). As another contribution, real images from different emotional classes are used in the generation process unlike the classical GAN approach which generates images from simple noise arrays. By using the proposed method, our system can classify emotions in facial images regardless of gender, race, ethnicity, age and face rotation. An accuracy of 75.2% was obtained on 7000 real images (14,000, also considering the generated images) from multiple combined facial datasets.

Keywords: generative adversarial network; emotion classification; facial key point detection; facial images processing; convolutional neural networks

---

1. Introduction

Face detection and recognition has been an on-going research area for the last 50 years, with concluding results being obtained starting with the late 90s [1]. The fast development of facial recognition technology allowed it to be used in a variety of areas like assisted living, health monitoring, access control, authentication, ID/passport control and fraud prevention, security/law enforcement (to identify lawbreakers or terrorists), surveillance systems, attendance tracking and counting and many others. According to a report published by MarketsandMarkets in 2017 [2], the global facial recognition market was estimated at 3.37 billion USD in 2016 and it is expected to grow up to 7.76 billion USD by 2022, with an annual growth rate of 13.9%.

Various methods have been used for facial detection and localization, and reviews of those methods are presented in References [3–5]. Different methods vary from template matching and knowledge-based methods to support vector machines, hidden Markov models and principal component analysis. The reviews concluded that the obtained accuracies for detection kept improving with each new method, but the selected samples for research were limited and had little variety, with good accuracies being obtained only on specific datasets. Neural networks-based face recognition improved the results of all previous methods and also brought an increase in efficiency and execution time. A variety of reviews [6–14] compare the advantages, disadvantages and results of multiple different neural network methods. The reviews mark the importance of CNNs (convolutional neural networks) and deep learning in the area of facial recognition, deep learning specifically being considered a huge step in the evolution of facial recognition algorithms. Most of the presented researches have accuracies over 90% on public available datasets, but different challenges are still acknowledged regarding real-world facial recognition, training the algorithms to replicate human behavior and large scale adoption in the industry. Different approaches are presented in References [15,16], where fuzzy algorithms perform a rotation invariant face recognition based on symmetrical facial characteristics. The main advantage is that the algorithms can be used on smart TVs (Television sets) with low processing power to recognize the viewer and offer proper content and services accordingly. The algorithm presented in Reference [16] is an enhanced version of the one in Reference [15], with an increase in accuracy. The presence of cosmetics and contact lenses adds challenges to face recognition for biometric purposes. Color, shape and texture features of the face and iris are extracted in Reference [17] to be used in a SVM (support vector machine) classifier for face recognition regardless of the makeup. The research shows improvement over several other face recognition methodologies. Another method was also developed in Reference [18], for makeup-invariant face verification, making use of the generative adversarial network (GAN) architecture first introduced in Reference [19]. The algorithm synthesizes non-makeup images from makeup images so that they can be used for face verification. The algorithm outperforms competing algorithms in terms of accuracy, speed, and size of the training dataset.

The introduction of GAN in Reference [19] opened new possibilities for image generation algorithms [20], including facial images. In this case, the generator (G) component is used to synthesize new images, while the discriminator (D) should detect the fake generated images. The G and D learn to improve by playing a minimax game which each of the components tries to win. There are two possible outcomes when using and training GANs. If more focus is put on the generator, then an image synthesis system is obtained. Otherwise, if the generator is used only to create images for the discriminator to assess, the D component can be used as a classifier. In Reference [21], a conditional GAN is used to generate facial images from simple noise and conditional data. This extension of the basic GAN is the first GAN model used strictly for facial generation. GANs can also be used to synthesize an aged version of the input image, as seen in Reference [22]. Although the results can't be validated, the obtained images are highly realistic. Other use cases for GAN include generating front-faced images from rotated images [23], altering images (closing/opening eyes/mouth) while preserving identity of the person illustrated in the images [24], and also removing extra lighting from facial images to ensure proper conditions for face identification [25]. The last three techniques prove the utility of GANs in image processing. The generator is trained in [26] to reconstruct 2.5-D images from 2-D images, and the output is used in two other CNNs (convolutional neural networks) for feature extraction and face recognition. Different training techniques for GANs are presented in References [27–31], covering unsupervised, semi-supervised, and supervised learning and also providing different outputs for classifiers:

- Class conditional models: condition the G to produce an image in a specific class and use the D to assert whether the image is fake or real (two output classes)
- N-output classes [27]: Use the D to classify the input image in various classes; ideally, the generated images should have a low level of confidence for the output class. The semi-supervised learning approach almost leads to the best performance in classifying images containing numbers or different objects. Unfortunately, the unsupervised approach has proven a weak accuracy in multiple-class classification.
- N+1-output classes [29–31]: Use the N-classes approach but also have a distinct class for generated images. The semi-supervised trained classifier in Reference [29] is a more data-efficient version of the regular GAN, delivering higher quality and requiring less training time. The research has been conducted on the MNIST database (Modified National Institute of Standards and Technology

database). The same conclusion was also reached in References [30] and [31] by the creators of the original GAN, but with an expanded dataset containing images of different objects, animals and plants.

2. Related Work

Emotion recognition is a new sub-area of facial recognition with high potential. Applications that perform emotion recognition can be used in various areas, like marketing (products/services evaluation and feedback based on customer emotions), psychology (identifying criminal profiles or terrorists before committing an attack), security (replace the panic button with fear detection during a robbery or an assault), and even medicine [32–35] (effects of positive and negative emotions on the patients' health using current technology). Although performed before the development of modern emotion recognition techniques, the presented medical studies show the importance of emotion monitoring as a step in detecting depression and other diseases. Most progress in using GANs in the domain of emotion is represented by the possibility of altering an emotion in an image based on labeled information about the target emotion [36–41]. The obtained images are highly realistic and hard to distinguish as fake by human observers. The method in Reference [36] and its improved version [41] generate a sketch image of the emotion from an image, its emotion label and random noise. The sketch is assessed by the discriminator for correctness and then used as input in another GAN which generates an image of another person with the same facial emotion. The generated facial expressions are compared with real valid facial expressions, having the distances between the two classes reported as small.

A starting point in emotion recognition is represented by the identification of facial regions of interest, which can be done by localizing a series of facial key points. These features describe the position, shape, and size of the corresponding regions of interest. In Reference [42], a lip contour detection and tracking system is presented. The system uses a multi-state mouth model that represents different mouth states, a series of lip templates, and shape, color and motion information. The facial points associated with the lip are tracked in the image sequence and the lip contour is obtained from the template parameters, with the color and shape information being used to distinguish different lip states. A neural network for the detection of 15 facial key points is described in Reference [43]. The proposed deep convolutional neural network uses a learning model for each facial key point with the result outperforming other similar approaches. A total of 194 facial landmarks are estimated for each facial image in Reference [44] by using an ensemble of regression trees. The obtained predictions are of high quality, with the algorithm also performing in real-time. The paper also includes optimizations for improving feature selection, a comparison of different regularization strategies, and a study on the evolution of predictions based on the quantity of training data. Facial micro-expressions are analyzed in Reference [45] using 31 facial points out of the 121 obtained using the Kinect face tracking API (Application Programming Interface). The micro expressions are analyzed based on different visual and auditory stimuli, as well as the gender of the subjects. The authors also studied the possibility to distinguish emotions based on the results.

Two different neural networks for emotion recognition are trained and compared in Reference [46]. The first approach is to use representational autoencoder units. Four autoencoders were developed and tested on the JAFFE (Japanese Female Facial Expression) [47] and LFW (Labeled Faces in the Wild) [48] facial images datasets with accuracies of 60% and 49% respectively. The other selected implementation is an eight-layer convolutional neural network, created and trained from scratch. The network includes convolutional, max pooling, and fully connected layers. Using the same datasets [47,48], the accuracy increased to 86% and 67%, respectively, after 20 epochs and 420 iterations. In Reference [49] a CNN classifier is developed and trained on the FER2013 dataset [50]. Due to differences in the number of images for each emotion class, two cycle-GANs are trained to generate disgust and sadness images starting from neutral face images. Therefore, the training dataset is expanded for an equal distribution of images. Using the generated images, the overall accuracy of the CNN classifier improved. Further testing with good results is performed on other datasets [47,48,50]. A fear estimation system is developed in Reference [51], using two images captured by a dual camera system: a near infrared (NIR) camera (Logitech, CA, USA) and a thermal camera (FLIR, OR, USA). Seven different features are extracted from the two images (two from thermal images and five from NIR images) and the last feature is represented by the direct input of the study subjects via a real time questionnaire. The algorithm proposed in [44] is used to extract 68 facial feature points for the NIR images. The extracted feature points are further used to compute the five features based on facial point movement between successive images of the subject who switches from neutral to scared (fear). The top four discriminatory features are selected and their values are normalized (0–1) and used as input in a fuzzy inference system, which evaluates the value of the fear emotion from low to high.

The authors in Reference [52] develop and train two convolutional neural networks with different scale invariant features. The feature descriptors are represented by image gradients computed using key points neighboring pixels of the given image, on 4×4 patches (16 patches for each image). K-means clustering is used to group the feature descriptors in clusters for each emotion. The proposed models are trained on FER [50] and CK+ [53] datasets and tested on an additional dataset, SFEW [54]. The reported results have a good accuracy on the training dataset, but a decreased one for the third dataset. In Reference [55], two methods for emotion recognition are proposed: SVM and CNN. The different SVM models (one-vs-one, principal component analysis, one-vs-all, histogram of oriented gradients) presented issues during training and obtained lower accuracies on all the tested datasets. Several other CNN implementations with additional preprocessing techniques were tested. The best obtained accuracy on a small subset of FER [50] was 66.67%. The algorithm is further used for real-time image classification in video feeds. Five existing CNN approaches for deep learning are proposed, adjusted, and compared in [56], with the scope of emotion recognition. The input images are preprocessed using the Viola-Jones algorithm. Then, existing models are adjusted (adding new layers), trained and tested for accuracy. A CNN with two similar sequences of two convolutional layers and a sub-sampling layer, followed by a dense layer with 3072 filters and an output layer, obtained the best accuracy (63%).

The current paper proposes a new system for emotion classification based on a GAN classifier. The facial emotions are classified within seven emotions–anger, disgust, fear, happiness, neutral, sadness, and surprise. To this end, 14 classes are used to train the GAN–a real class and a fake one of each emotion. The novelty of the proposed method is brought by using the new 2N-classes approach for training the GAN classifier which normally operates with N-classes. As a consequence, the detection accuracy increased. Another contribution is the expansion of the test images dataset by generating images using the GAN. Real images of a different class are used in the generation process, which is different from the standard GAN approach to generate images from a simple noise array. By only using the rotation-invariant facial points as input for the classifier, we also reduce the amount of data that is analyzed. The facial-points vectors are processed to be rotation insensitive, so that tilted facial images can also be classified, as opposed to similar presented algorithms, which can classify only front faced facial images. The remainder of the paper is organized as follows: In Section 3, the methodology and architecture of the proposed system are described. In Section 4, the experimental results are presented, along with a performance analysis. The paper concludes with the discussions in Sections 5 and 6.

3. Materials and Methods

3.1. Training and Evaluation Phase

3.1.1. System Architecture

Robert Plutchik [57] developed a wheel of emotions, stating that there are eight primary emotions: happiness (joy), sadness, anger, fear, trust, disgust, surprise, and anticipation, which can have a variety of intensities. The primary emotions are located on the first ring. Moreover, complex emotions can be obtained from a mix of primary emotions (with a distance of 1, 2 or 3 on the wheel), thus obtaining the full spectrum of human emotions.

We propose a system for the classification of six primary emotions (happiness, sadness, anger, fear, disgust, and surprise) in facial images, adding another class of neutral emotion (lack of a dominant emotion). Five emotions are negative, with happiness being the only positive. The system is based on a modified conditional GAN. The first proposed implementation of a GAN [19] had a simple structure. The discriminator D would receive either a real image or a fake (generated) image and would have to assess it as real or fake. The generator G was responsible with generating a fake image similar to the real one, starting from simple noise and a latent space vector. Based on the correctness of the decision, the discriminator and generator would adjust their weights. The discriminator and generator play a minimax two-player game with the value function in Equation (1):

$$\min_{G}\max_{D} V(D,G) = E_{x\sim p_{data}(x)}[\log D(x)] + E_{z\sim p_z(z)}[\log(1-D(G(z)))] \quad (1)$$

The first term of the equation is represented by the entropy (E) passed by the distribution of the real data ($p_{data}(x)$) through the discriminator (D(x)) and it can have a maximum value of 1. The second term is represented by the entropy passed by the distribution of the random noise input ($p(z)$) through the generator (G(z)) that produces a fake data sample which is further passed to the discriminator for assessment. The second term can have a maximum value of 0. The discriminator tries to maximize the value function V(D,G) (meaning that the fake data is always labeled as fake), while the generator tries to minimize the value function (in this case the difference between the real and the fake data is minimum)

Starting from the original network structure, several varieties of GAN architectures were proposed, as seen in Figure 1.

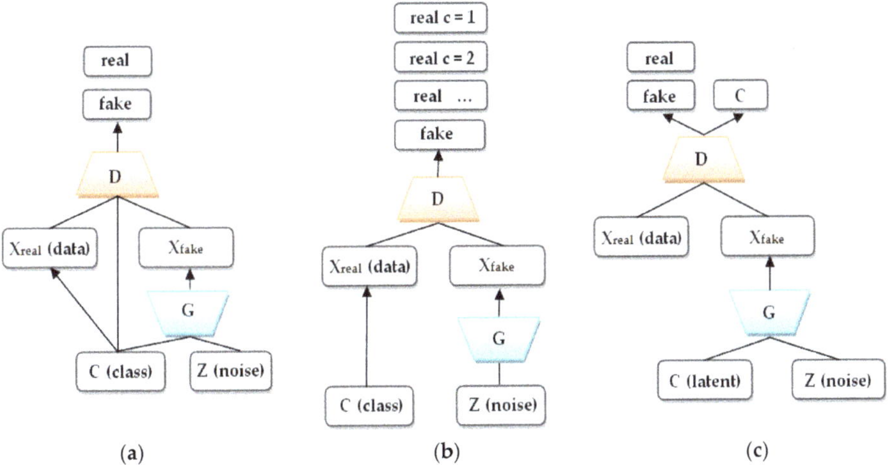

Figure 1. Different GAN (generative adversarial network) implementations: (**a**) Conditional GAN [19]; (**b**) Semi-Supervised GAN [29]; (**c**) Info-GAN [58].

Our proposed architecture combines elements from the previous described implementations. The novelty is brought by using a real image not part of the desired class to generate the fake images, instead of using a noise vector, adding an image processing block for facial points extraction and constructing rotation invariant facial vectors, and splitting the real/fake assessment and class identification into a single 2N type classification (a real and a fake class for each emotion). The proposed architecture can be seen in Figure 2.

Each training cycle of the network is split into three phases. During the first phase (flow I—the left side of Figure 2), the generator is switched off and the discriminator receives only real class-labelled images. The discriminator adjusts its weights based on the feedback loop FD. For the first phase of the first training cycle, the discriminator will only use the N real classes as possible outputs for an image. For any other phase or cycle, all the 2N classes are used. During the second phase (flow II- the right side of Figure 2), the discriminator remains unmodified and the generator is trained to deliver fake images of given classes which the discriminator has to classify. The generator uses the feedback loop FG to adjust weights. In the third phase (also flow II), the roles switch and the generator is kept unmodified, while the discriminator is trained with both real and fake images. The feedback loop FD is used for weights adjusting. The three main components (image processing block, discriminator and generator) are described in Section 3.1.2, Section 3.1.3, and Section 3.1.4, respectively.

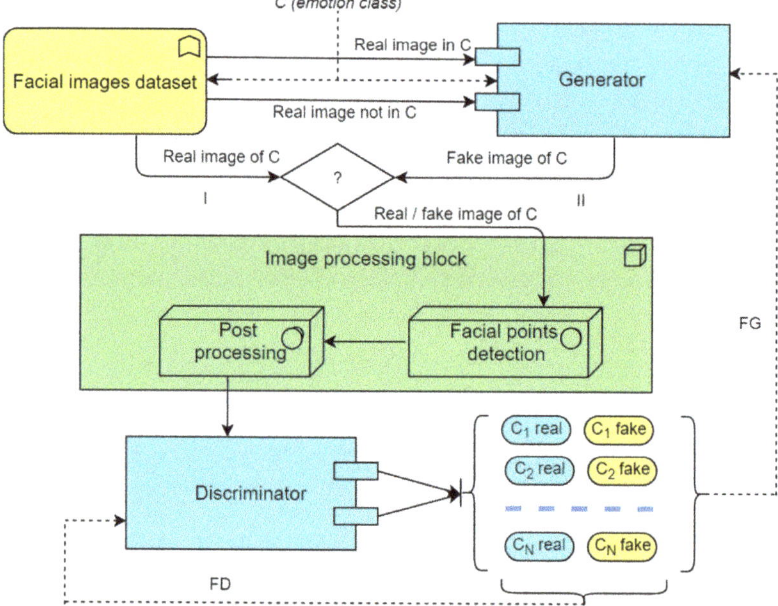

Figure 2. Proposed GAN architecture.

3.1.2. Image Processing Block

The image processing block acts as an intermediate between the input images (either real or generated) and the discriminator. We designed this block so that the discriminator can receive more meaningful information based on which it can classify the images. This block performs two main operations, namely the detection of facial-key points (detailed in Section A) and finding the correlation between these points (Section B). The image processing block is used to minimize the variations brought by gender, age, race, and head posture, while using a large range of test images. Similar works try to limit these variations by limiting the image dataset on which the algorithms are validated.

A. Facial Points Detection

Facial landmarks are regions of interest that can uniquely identify different components of the face, such as eyes, eyebrows, lips and nose. These landmarks can be described by a series of facial key points. In order to extract the facial feature points, we used the real-time face estimation open source code from dlib C++ library [59]. The code implements the method described in Reference [44]. The dlib library contains a pre-trained detector that estimates the coordinates of 68 points that are mapped on facial regions of interest. The implemented detector uses an ensemble of regression trees for facial feature tracking. The 68 labeled points output of the detector can be seen in Figure 3a, while Figure 3b,c show the result of applying the detection algorithm on a test image. Because most of the test images only contain a cropped image of the face, we will not use the full set of 68 points, but a smaller one of 51 (removing the 17 points associated with the jaw line).

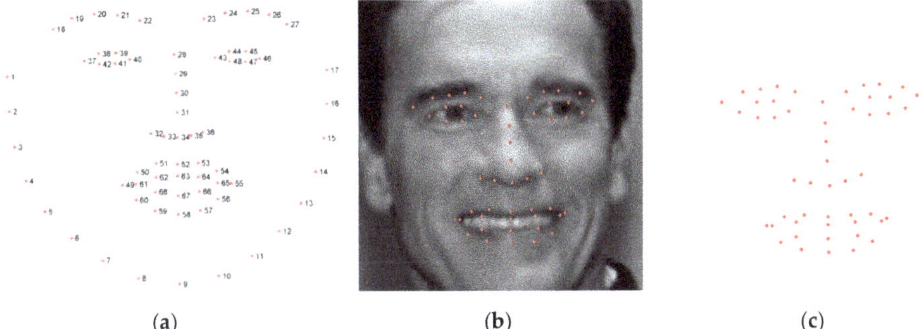

(**a**) (**b**) (**c**)

Figure 3. Images resulted from dlib detector (**a**) 68 points; (**b**) Initial image with facial key points; (**c**) 51 extracted facial key points.

The facial regions of interest can be described as follows (using the points from Figure 3a):

- Right eyebrow—points 18, 19, 20, 21 and 22;
- Left eyebrow—points 23, 24, 25, 26 and 27;
- Right eye—points 37, 38, 39, 40, 41 and 42;
- Left eye—points 43, 44, 45,46, 47 and 48;
- Nose—points 28, 29, 30, 31, 32, 33, 34, 35 and 36;
- Mouth:
    - Upper outer lip—points 49, 50, 51, 52, 53, 54, and 55;
    - Upper inner lip—points 61, 62, 63, 64, and 65;
    - Lower inner lip—points 61, 65, 66, 67, and 68;
    - Lower outer lip—points 49, 55, 56, 57, 58, 59, and 60.

B. Post Processing

In this module, we computed the relative position of the facial points relative to each other. In order to achieve this, we first computed the position of the facial center of gravity as the average position of all the other extracted points from Section A, using the Equation (2), where x_i represents the distance on the OX axis and y_i represents the distance on OY axis, from the center of origin O located in the lower left corner of the image.

$$x_{mean} = \frac{\sum_{i=18}^{68} x_i}{51} \qquad y_{mean} = \frac{\sum_{i=18}^{68} y_i}{51} \tag{2}$$

After determining the center of gravity, we computed the vectors that join the center of gravity and the other facial key points. Each of the vectors has a direction (angle relative to the horizontal axis) and a magnitude (distance from the center of gravity). In Figure 4, the center of gravity (blue dot), the facial key points (red dots) and the vectors connecting them (green lines) can be observed. Also, symmetry between vectors corresponding to the same points on the left and right sides of the face can be observed.

The center of gravity was selected as reference over any of the points because of the variance the different points bring depending on the face morphology. This method did not completely solve the variance brought by the rotation of the face relative to the camera around the vertical (OY) or horizontal axes (OX). For the scope of this paper, only the rotation along the third axis (OZ, head tilt) will be corrected. During the initial pre-research that was performed to study the feasibility of the proposed method, we identified that other similar works used only front-faced non-rotated facial images. The possibility of classifying tilted facial images was investigated. By using the initial obtained facial vectors of the tilted images, the resulting classification accuracy of these images was low. By reducing the distance between the front-faced posed vectors and the tilted vectors, we managed to match the accuracy between the two situations. For this purpose, the angular offset β between the line obtained by joining points (28, 29, 30, 31 and 34) and the vertical axis (parallel with OY) starting from point 34 was computed. The angle β showed the tilt that should be corrected. Using this offset, the obtained vectors could be rotated so that the faces have a uniform (front-facing) pose, while keeping the same expression. For each vector, the new direction angle γ and new positions x' and y' are computed as in Equation (3), with α being the original angle formed by the vector with the OX axis in the tilted image and x and y the original positions:

$$
\begin{aligned}
\alpha &= \tan^{-1}\left(\frac{y-y_{mean}}{x-x_{mean}}\right) \times \frac{180}{\pi} \\
\beta &= \tan^{-1}\left(\frac{x_{28}-x_{34}}{y_{28}-y_{34}}\right) \times \frac{180}{\pi} \\
\gamma &= \alpha + \beta \\
x' &= x_{mean} + \cos(\gamma)\sqrt{(x-x_{mean})^2 + (y-y_{mean})^2} \\
y' &= y_{mean} + \sin(\gamma)\sqrt{(x-x_{mean})^2 + (y-y_{mean})^2}
\end{aligned}
\tag{3}
$$

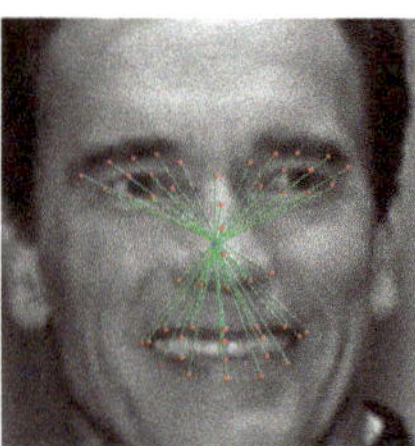

Figure 4. Center of gravity and connections with facial key points.

The visual interpretation of the above described procedure can be seen in Figure 5:

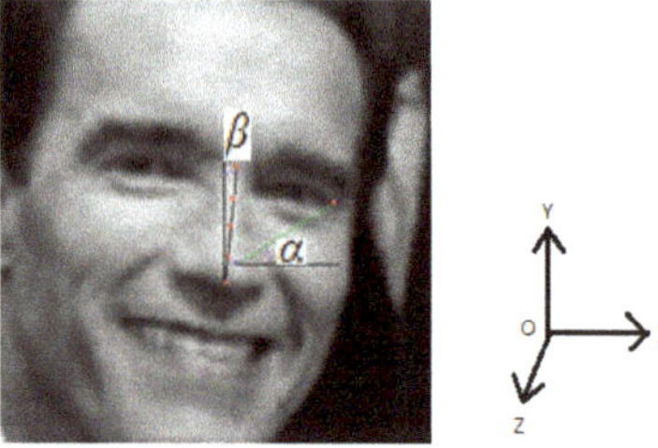

Figure 5. Computing the offset to correct face tilt.

3.1.3. Discriminator

The proposed CNN structure for the discriminator consists of three convolutional layers, three pooling layers (two max-pooling and one average-pooling), two fully-connected layers and an output Softmax layer. The architecture is presented in Figure 6.

The input is represented by a 48 × 48 pixels grayscale image. Each of the three convolutional layers use 3 × 3 filter functions, with a stride of 1 and a padding of 1. The 0-padding was used to maintain the size of the output feature maps. The number of convolution filters increases from 32 (convolution layer 1) to 64 (convolution layer 2), and 128 (convolution layer 3), respectively. Each convolution layer is followed by a pooling layer. All three pooling layers which are used (one average-pooling and two max-pooling) have a stride size of 2 × 2 and dropouts of 0.1. The final two fully connected layers use 256 and 128 neurons, respectively, with dropouts of 0.4 and 0.5. The final layer of the proposed CNN is a Softmax layer with 14 possible outputs (7 emotion classes and real/fake classification).

The discriminator neural network was developed using Python and the machine learning framework, Tensorflow. It uses a new 2N output classes approach, by having a real and a fake class for each emotion. This approach helped improve the overall emotion classification by having the discriminator also trying to associate fake images with emotion classes of interest, instead of just rejecting the images as fake (N+1-classes approach).

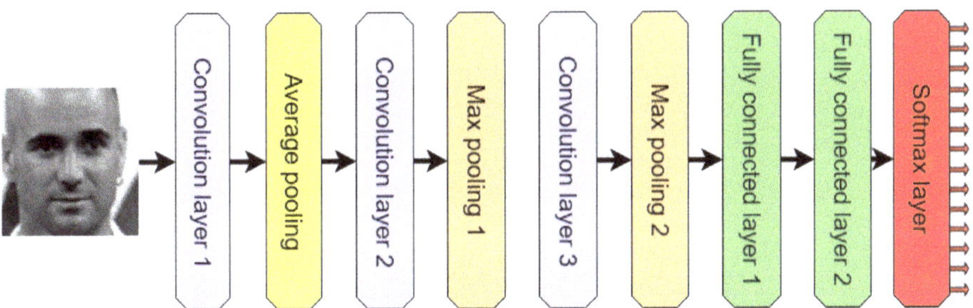

Figure 6. Discriminator architecture.

3.1.4. Generator

The generator performs realistic facial expression synthesis. It receives a facial image that has to be modified, the target expression, and a sample facial image of the target expression, and then generates an image of the initial person with the expression of the second person, defined by the target emotion. The initial and generated images are 48 × 48 pixels grayscale images ($R^{48 \times 48}$). Both the initial (I) and the label image (I_L) are processed by a four convolutional-layer network (encoder Enc_i), the initial image being mapped to a latent vector and the label image to a label vector, respectively. The concatenation result of the two vectors is used by a four deconvolutional-layer network (decoder-*Dec*) to generate the target image ($\tilde{I}$). The fully connected layer of the decoder learns the differences between the two vectors (latent-initial image and label-target/label image). The feedforward loop (*FFL*) is used to provide the raw features of the initial image (a down sampled version of the initial image), on which the differences identified by the first six layers of the decoder is applied. The formula for the obtained image is presented in Equation (4):

$$\tilde{I} = Dec(Enc_1(I), Enc_2(I_L),\ FFL) \tag{4}$$

The description of the used layers is:

- Convolutional layers (1a–4a, 1b–4b)
    - 5 × 5 filter functions, stride 1, padding 2 (0-padding)

  - Layers 1 and 2–128 neurons, Layers 3 and 4–256 neurons
- Max pooling layers (1a–4a, 1b–4b) with stride 2 × 2
- Fully connected layers
  - 256 neurons for the encoders, 512 for the decoder
- Deconvolutional (transposed convolution) layers (1–4)
  - 5 × 5 filter functions, stride 1, padding 2 (0-padding)
  - Layers 1 and 2–128 neurons, Layers 3 and 4–256 neurons
- Upsampling layers (1–4) with stride 2 × 2
- Leaky ReLU as activation function—gradient 0.15

In most GAN implementation, a continuous noise vector is used to generate the images. The noise vector has no actual relevant information, but it is a source of randomness. By processing an initial image that has to be converted to a different facial expression, along with another image that has the desired facial expression, we construct a meaningful vector that is further used in the emotion-guided image generation process.

The generator neural network system was developed using Python and the machine learning framework, Tensorflow. The architecture is presented in Figure 7.

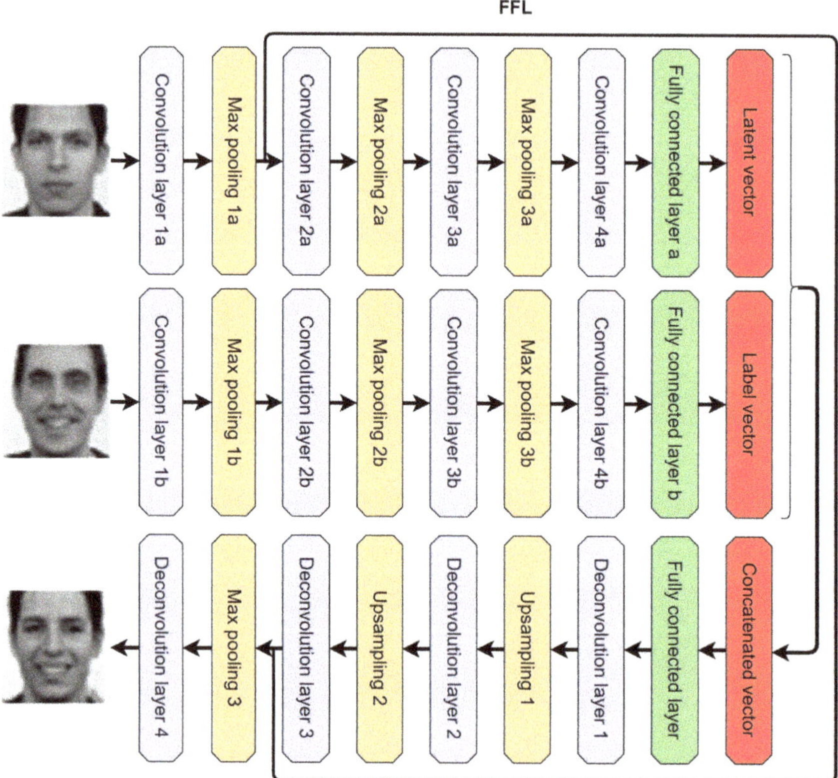

Figure 7. Generator architecture.

3.2. Operational Phase

After the proposed implementation in Figure 2 is trained and validated, several changes are made for the system to run independently. The classification part is the main component of the new system. There are three major changes from the proposed implementation in Figure 2. Firstly, the input image is provided by the user. The input image has to be a facial 48 × 48 pixels grayscale image. For the scope of this paper this is a mandatory requirement, but, for a future implementation, we consider adding another processing block so that the user can input a different size image and it will be converted to 48 × 48 pixels grayscale facial image. The second change is that the real and fake classes for each emotion are merged into a single class for each emotion. Both real and fake classes of the same emotions are considered to be the same class in this phase. This division was originally done during the training phase to increase the accuracy of the system. Thus only seven output classes remain. Finally, the feedback loop that was used to adjust the discriminator weights during supervised learning is removed, due to the fact that the images in this phase are not labeled. The new architecture for the classification system can be seen in Figure 8a. In order to reuse the generator, an additional system is proposed. The user can input a 48 × 48 pixels grayscale facial image and a target emotion for it, and the selected emotion will be transferred to the input image, changing the facial expression accordingly. The generator uses a random image belonging to the selected emotion class from the initial labeled dataset that was used for training. The proposed architecture for the generator system can be seen in Figure 8b.

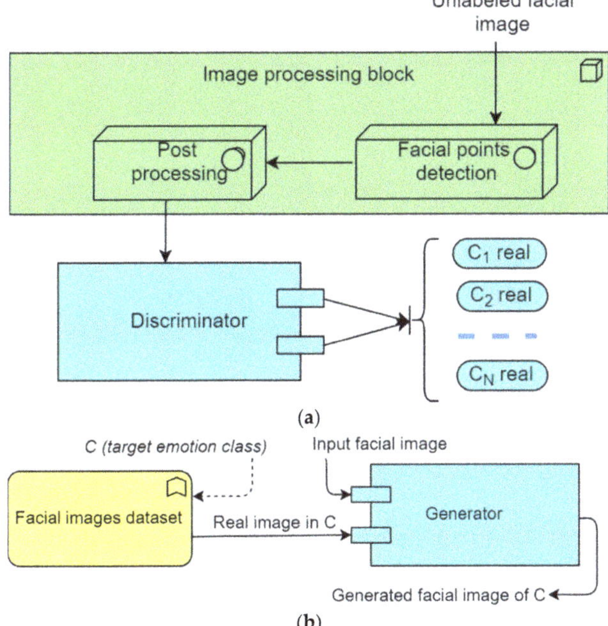

Figure 8. Discriminator and generator adapted for the operational phase: (**a**) Discriminator; (**b**) Generator.

4. Experimental Results

In order to train the proposed classification system we selected 7000 images (1000 images for each emotion class) from multiple datasets: LFW [48], FER 2013 [50], CK+ [53] and SFEW [54], FER+ [55]. Around 85% of the 7000 images used in this phase were selected from the FER 2013 dataset, which has the greatest diversity of the mentioned datasets, also being one of the largest open-source datasets for emotion recognition (almost 30,000 labeled images). The FER 2013 dataset consists of pre-cropped grayscale images of size 48 × 48, so all other selected images from different smaller datasets were manually cropped to have the same face pose and converted from RGB to grayscale. By using images from different datasets, we added an additional variety that the system had to handle. A selection of images for each emotion class can be seen in Figure 9.

The proposed system was implemented using Python and the Tensorflow machine learning framework. The algorithm was tested on system with 32GB DDR4 and a NVIDIA GeForce GTX 950M GPU with 4GB dedicated GDDR5 memory (NVIDIA Corporation, Santa Clara, CA, USA). For this setup we made use of the Tensorflow-CUDA (Compute Unified Device Architecture) toolkit integration, to enable parallel computing and obtain better execution times and performance. The system was trained for 200 epochs, when it was observed that the accuracy did not significantly improve anymore. Each epoch consisted of two sub-epochs. During the first sub-epoch, all 7000 test images are passed to the discriminator for classification (left side in Figure 2-I.). The image dataset was randomly split into two equal parts in sub-epoch 2 (right side in Figure 2-II). The first 3500 images were used to train the generator with the discriminator kept unchanged, while the next 3500 were used to test the discriminator with the generator unmodified. The batch size was 100 images in all scenarios, thus having 70 iterations for each sub-epoch and 140 iterations per epoch. The execution time averaged out at 6 hours per epoch (2 h for the first sub-epoch and 4 h for the second one).

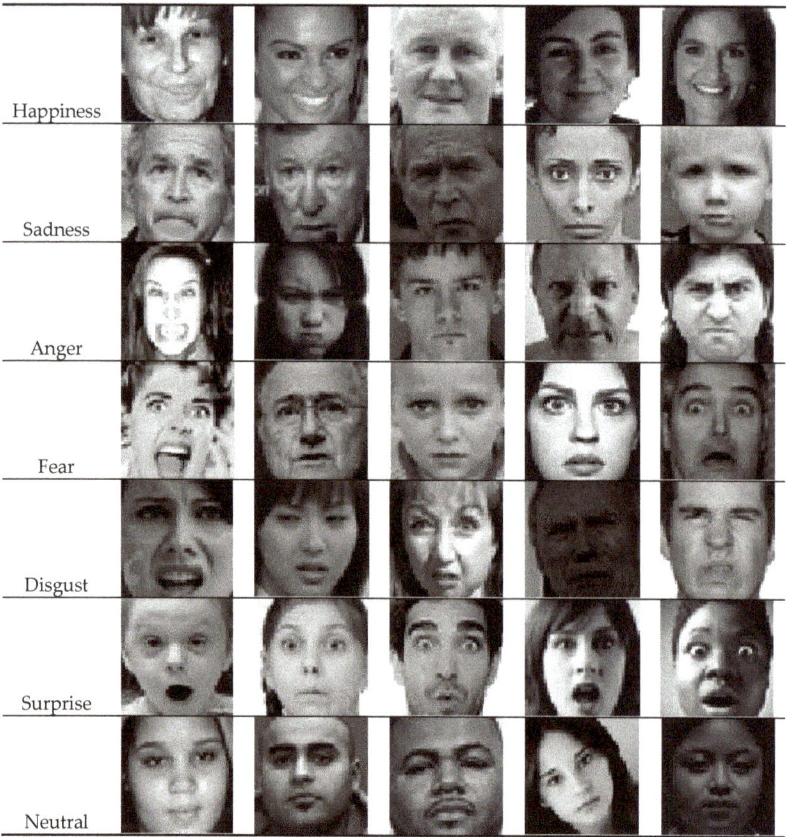

Figure 9. Sample images from the selected datasets.

In Figure 10, original, labeled and generated labeled images can be observed. These images are obtained using the right path in Figure 2-II (right side of Figure 2). After the training phase (200 epochs), a different set of 7000 images was selected from the FER 2013 dataset.

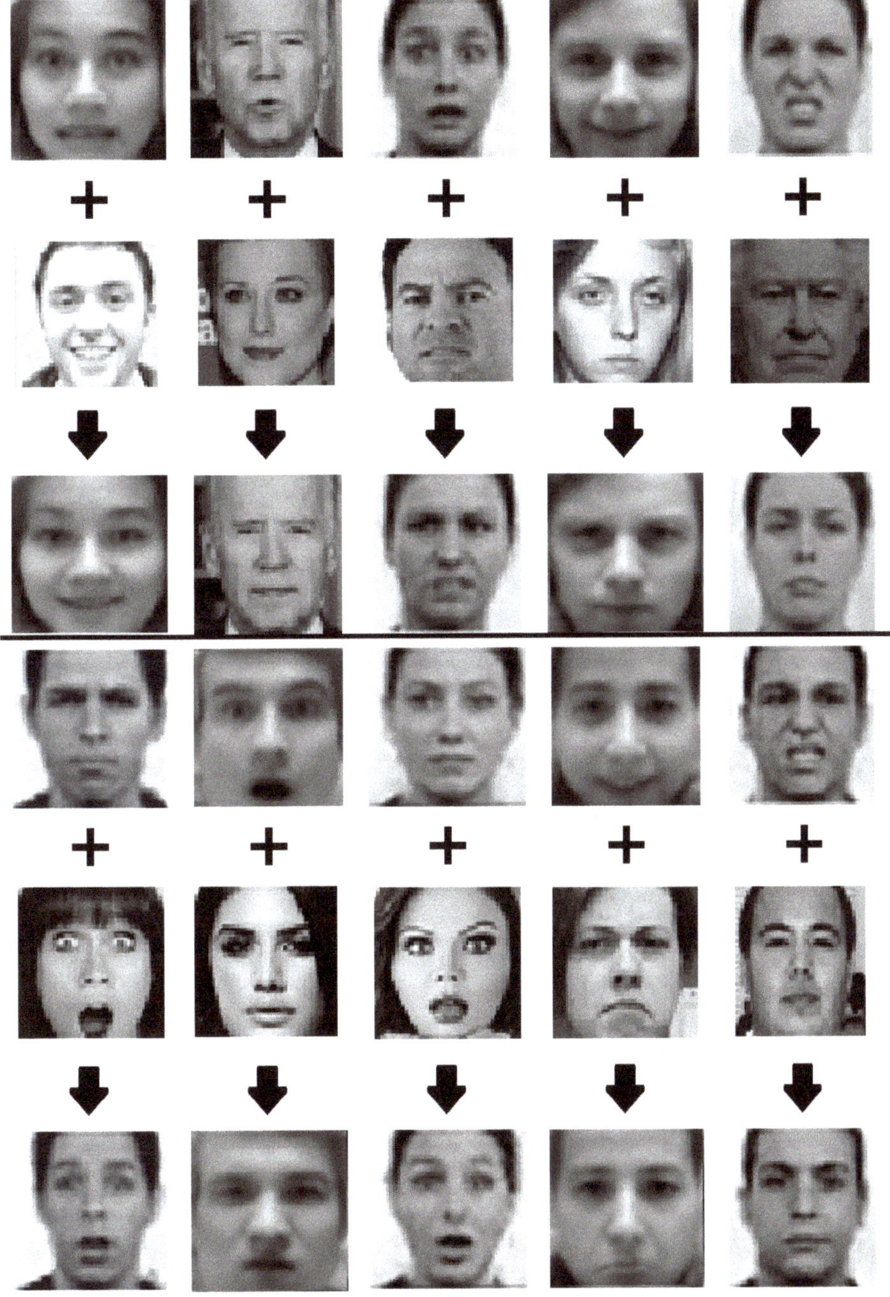

Figure 10. Generator results.

The proposed classification system was retested for another epoch using the new dataset. In Table 1, the confusion matrix obtained during the last epoch can be seen. For each emotion there were 2000 images, half real(R) and half generated (fake, F), like in each of the training epochs. The true positive entities were highlighted with gray.

In order to assess the performance of the proposed system we considered as a starting point the well-known statistical terminology:

- *TP*–Number of true positives, positive correctly classified as positive
- *TN*–Number of true negatives, negative correctly classified as negative
- *FP*–Number of false positives, negative classified as positive
- *FN*–Number of false negatives, positive classified as negative

We further compute statistical measures using the values described above. The measures and their formulas can be observed in Table 2:

Table 1. Confusion matrix of the proposed system.

		Happiness (H)		Sadness (SA)		Anger (A)		Fear (FE)		Disgust (D)		Surprise (SU)		Neutral (N)	
		R	F	R	F	R	F	R	F	R	F	R	F	R	F
H	R	911	37	0	0	0	0	0	0	0	0	19	7	21	5
	F	27	923	0	0	0	0	0	0	0	0	2	18	7	23
SA	R	6	0	704	62	19	3	9	0	55	9	3	0	97	33
	F	0	3	57	705	2	11	0	4	7	63	0	1	37	110
A	R	2	0	5	0	765	65	23	6	54	14	39	15	12	0
	F	0	1	0	2	92	771	5	17	12	50	14	25	2	9
FE	R	3	0	14	3	21	5	715	83	13	4	62	21	44	12
	F	0	0	4	10	5	10	89	719	2	7	19	61	13	51
D	R	2	0	15	4	67	20	15	4	753	82	9	3	21	5
	F	0	0	1	14	19	65	3	18	78	767	3	11	3	18
SU	R	78	28	0	0	3	0	58	15	13	3	712	85	5	0
	F	22	77	0	0	0	7	18	61	2	17	88	705	1	2
N	R	74	22	63	15	14	1	17	3	22	3	22	1	683	60
	F	12	76	10	52	0	12	1	18	4	21	2	32	64	696

Table 2. Statistical measures of performance.

True positive rate (TPR)/sensitivity	$\frac{TP}{TP+FN}$	False positive rate (FPR)	$\frac{FP}{FP+TN}$
True negative rate (TNR)/specificity	$\frac{TN}{TN+FP}$	False negative rate (FNR)	$\frac{FN}{TP+FN}$
Positive prediction value (PPV)/precision	$\frac{TP}{TP+FP}$	False discovery rate (FDR)	$\frac{FP}{TP+FP}$
Negative prediction value (NPV)	$\frac{TN}{TN+FN}$	Accuracy (ACC)	$\frac{TP+TN}{TP+FP+TN+FN}$

For each emotion class, we compute the statistical measures in each of the following cases:

- Only real images of the class considered as positive (R)
- Only fake images of the class considered as positives (F)
- All images of the class (both real and fake) considered as positive (R + F)
- Real/fake differentiation (real images are positive, fake images are negative, regardless of the class) (R/F)

The results can be seen in Table 3, with the same notation for each emotion class as in Table 1.

The overall accuracy of the proposed system was 75.2%, while the accuracy of distinguishing between true and generated images was 82.9% (highlighted with gray). This final test was repeated, but this time without the generator module and the seven fake output classes, which were disabled.

By doing this, we wanted to determine the improvement in accuracy brought by using the generator and the fake emotion classes. Only 7000 real images were used, and the obtained accuracy was 73.2%. Therefore, it was determined that adding the fake images in the classifications process contribute to a 2% increase in accuracy and variation of the tested images. There is no significant difference in accuracy between the tilted images and the front-faced images due to using the adjusting method presented in Section 3.1.2.

Table 3. Performance results.

		TPR	TNR	PPV	NPV	FPR	FNR	FDR	ACC
H	R	0.911	0.982	0.801	0.993	0.018	0.089	0.199	0.977
	F	0.923	0.981	0.791	0.994	0.019	0.077	0.209	0.977
	R + F	0.917	0.982	0.796	0.985	0.018	0.083	0.204	0.954
	R/F	0.951	0.964	0.963	0.951	0.036	0.049	0.037	0.9575
SA	R	0.704	0.987	0.806	0.977	0.013	0.296	0.194	0.966
	F	0.705	0.987	0.813	0.977	0.013	0.295	0.187	0.967
	R + F	0.704	0.987	0.809	0.951	0.013	0.296	0.191	0.934
	R/F	0.893	0.897	0.896	0.893	0.103	0.107	0.104	0.895
A	R	0.765	0.981	0.759	0.982	0.019	0.235	0.241	0.965
	F	0.771	0.984	0.794	0.982	0.016	0.229	0.206	0.969
	R + F	0.768	0.983	0.776	0.961	0.017	0.232	0.224	0.933
	R/F	0.900	0.875	0.878	0.897	0.125	0.100	0.122	0.887
FE	R	0.715	0.982	0.750	0.978	0.018	0.285	0.250	0.962
	F	0.719	0.982	0.758	0.978	0.018	0.281	0.242	0.963
	R + F	0.717	0.982	0.754	0.953	0.018	0.283	0.246	0.926
	R/F	0.872	0.869	0.868	0.87	0.131	0.128	0.132	0.865
D	R	0.753	0.980	0.741	0.980	0.020	0.247	0.259	0.963
	F	0.767	0.979	0.737	0.982	0.021	0.233	0.263	0.963
	R + F	0.760	0.979	0.739	0.958	0.021	0.240	0.261	0.927
	R/F	0.882	0.893	0.891	0.883	0.107	0.118	0.109	0.887
SU	R	0.712	0.978	0.716	0.977	0.022	0.288	0.284	0.959
	F	0.705	0.979	0.715	0.977	0.021	0.295	0.285	0.958
	R + F	0.709	0.978	0.716	0.951	0.022	0.291	0.284	0.918
	R/F	0.869	0.869	0.869	0.869	0.131	0.131	0.131	0.869
N	R	0.683	0.975	0.676	0.975	0.025	0.317	0.324	0.954
	F	0.696	0.975	0.679	0.976	0.025	0.304	0.321	0.954
	R + F	0.690	0.975	0.678	0.948	0.025	0.310	0.322	0.908
	R/F	0.895	0.907	0.905	0.896	0.093	0.105	0.095	0.901
Total	R	0.749	0.981	0.750	0.980	0.019	0.251	0.250	0.749
	F	0.755	0.981	0.754	0.981	0.019	0.245	0.246	0.755
	R + F	0.752	0.981	0.752	0.958	0.019	0.248	0.248	0.752
	R/F	0.894	0.896	0.895	0.894	0.104	0.106	0.105	0.829

5. Discussion

A great variety of images (more than 14,000) from five different datasets (FER, FER+, LFW, CK+, SFEW) was used to test and validate the proposed system. The differences brought by gender, race, ethnicity, or age are minimized by computing the facial key points and the facial vectors from the center of gravity. By using this approach, we also handled the errors brought by tilted facial images, by adjusting the direction and magnitude of the facial vectors based on the face rotation. During the learning phase, the proposed CNN for emotion classification was tested both with real and generated images (thus increasing the variety to 28,000 images). Using the GAN approach to also generate images helps extend the available dataset and also introduces a greater variety of images. During each training epoch, the weights of the discriminator and generator are adjusted accordingly. This implementation increased the overall individual accuracy for each emotion class (R + F as opposed to R only), as can be seen in Table 3. It can be noted that the individual accuracy (class vs non-class) was quite high for each of the seven classes, ranging from 90% (neutral) to 97% (happiness). This variation can be explained by the fact that happiness was the only positive emotion we tested and can be easily distinguishable from the negative emotions. The lack of any emotion (neutral) was the closest to any emotion class and therefore more difficult to distinguish. Statistical comparison with similar works validated the proposed system, as observed in Table 4. In order to properly compare the results, we retested the algorithm for each distinct dataset (as opposed to the learning phase, where we used a selection of images from multiple datasets).

Table 4. Accuracy (%) comparison for emotion classification.

Dataset	[46]	[49]	[50]	[52]	[55]	[56]	Our Method
FER 2013	-	94.7	71	73.4	63	66.7	75.2
CK+	-	-	-	99.1	-	98.4	98.3
SFEW	-	39	-	52	-	-	60.8
LFW	67.7	-	-	-	-	-	75.7
JAFFE	86.4	95.8	-	-	-	-	94.8

It can be observed that the accuracy of our system is among the highest for the FER 2013 dataset. The most notable accuracy obtained on FER 2013 dataset was 94.7, but it was obtained on a small subset of images (the authors from Reference [49] reported using 7% and 14% of the images in the FER 2013 dataset, while in the current research, we used almost 50% of the available images. The reported results were slightly better when comparing the 7% case, with an overall accuracy and accuracy for five emotions being better, with the 14% case, where accuracy for two emotion classes was better). Although the FER 2013 images represented a great percent of the images used in the learning phase, the system was able to properly classify images from the other used datasets, as shown by the obtained accuracies in each of the respective cases. Finally, we tested the system on a new dataset, JAFFE [47], which was not used at all during the learning phase. Due to using the image processing block (facial points detection and post-processing) to minimize image variations, the system was able to correctly classify the new images with a high accuracy (94.8%).

6. Conclusions and Further Work

The proposed method, based on a Generative Adversarial Network, for emotion detection improved the classification accuracy for five combined facial dataset (75.2%—the overall accuracy, and 82.9%—the accuracy of identification true/generated images). The obtained system (operational phase) was flexible, allowing the use of images with great differences (gender, age, and race) as inputs. Moreover, the generator could be used as a standalone component for emotion change in any image. In order to reduce the calculus volume, the rotation-invariant facial points were used as inputs for the classifier of seven emotions.

One future research direction is represented by trying to identify a correlation between the emotions expressed by different individuals over a period of time and the evolution of their health state. This kind of study implies monitoring the persons at random intervals in their natural state using their smartphone, laptop, or smart TV camera and finding their predominant emotion in different situations throughout the day. The study is guided by the idea that a negative emotion can have impact on the overall health state, leading to stress and ultimately to diseases like cancer [32–35]. A strong collaboration with a medical institute is planned.

Another research direction is represented by the possibility to monitor and evaluate the emotion caused by different advertising campaigns (photos or videos) using the smartphone camera. In this way we can assess how well the campaign is received by the public.

Author Contributions: T.C. designed and implemented the classification system for learning, testing, and operational phase, and wrote the paper; D.P. came up with the concept and supervision, and L.I. selected and converted the test images to a standard format, and tested and validated the classification system.

Funding: This research was funded by University POLITEHNICA of Bucharest, grant number GEX 25/2017, CAMIA.

Acknowledgments: This work has been supported by University POLITEHNICA of Bucharest, through the "Excellence Research Grants" Program, UPB—GEX 2017. Identifier: UPB—GEX 22/2017, SET.

Conflicts of Interest: The authors declare no conflict of interest. The founding sponsors had no role in the design of the study; in the collection, analyses, or interpretation of data; in the writing of the manuscript; or in the decision to publish the results.

References

1. Wiskott, L.; Kruger, N.; Von Der Malsburg, C. Face recognition by elastic bunch graph matching. *IEEE Trans. Pattern Anal. Mach. Intell.* **1997**, *19*, 775–779. [CrossRef]
2. Face Recognition Market by Component, Technology, Use Case, End-User, and Region–Global Forecast to 2022. Available online: https://www.marketsandmarkets.com/Market-Reports/facial-recognition-market-995.html (accessed on 21 March 2018).
3. Yang, M.H.; Kriegman, D.J.; Ahuja, N. Detecting Faces in Images: A survey. *IEEE Trans. Pattern Anal. Mach. Intell.* **2002**, *19*, 775–779.
4. Gupta, V.; Sharma, D. A study of various face detection methods. *Int. J. Adv. Res. Comput. Commun. Eng.* **2014**, *3*, 6694–6697.
5. Hiyam, H.; Beiji, Z.; Majeed, R. A survey of feature base methods for human face detection. *Int. J. Control Autom.* **2015**, *8*, 61–77.
6. Smrti, T.; Nitin, M. Detection, segmentation and recognition of face and its features using neural network. *J. Biosens. Bioelectron.* **2016**, *7*. [CrossRef]
7. Le, T.H. Applying Artificial Neural Networks for Face Recognition. *Adv. Artif. Neural Syst.* **2011**. [CrossRef]
8. Farfade, S.S.; Saberian, M.; Li, L.J. Multiview face detection using deep convolutional neural networks. In Proceedings of the 5th International Conference on Multimedia Retrieval (ICMR), Shanghai, China, 23–26 June 2015.
9. Martinez-Gonzalez, A.N.; Ayala-Ramirez, V. Real time face detection using neural networks. In Proceedings of the 10th Mexican International Conference on Artificial Intelligence, Puebla, Mexico, 26 November–4 December 2011.
10. Kasar, M.M.; Bhattacharyya, D.; Kim, T.H. Face recognition using neural network: A review. *Int. J. Secur. Appl.* **2016**, *10*, 81–100. [CrossRef]
11. Al-Allaf, O.N. Review of face detection systems based artificial neural networks algorithms. *Int. J. Multimed. Appl.* **2014**, *6*. [CrossRef]
12. Prihasto, B.; Choirunnisa, S.; Nurdiansyah, M.I.; Mathulapragsan, S.; Chu, V.C.; Chen, S.H.; Wang, J.C. A survey of deep face recognition in the wild. In Proceedings of the 2016 International Conference on Orange Technologies, Melbourne, Australia, 17–20 December 2016. [CrossRef]
13. Fu, Z.P.; Zhang, Y.N.; Hou, H.Y. Survey of deep learning in face recognition. In Proceedings of the 2014 International Conference on Orange Technologies, Xi'an, China, 20–23 September 2014. [CrossRef]
14. Wang, M.; Deng, W. Deep face recognition: A survey. *arXiv* **2018**, arXiv:1804.06655.
15. Kim, Y.G.; Lee, W.O.; Kim, K.W.; Hong, H.G.; Park, K.R. Performance enhancement of face recognition in smart TV using symmetrical fuzzy-based quality assessment. *Symmetry* **2015**, *7*, 1475–1518. [CrossRef]
16. Hong, H.G.; Lee, W.O.; Kim, Y.G.; Kim, K.W.; Nguyen, D.T.; Park, K.R. Fuzzy system-based face detection robust to in-plane rotation based on symmetrical characteristics of a face. *Symmetry* **2016**, *8*, 75. [CrossRef]
17. Sharifi, O.; Eskandari, M. Cosmetic Detection framework for face and iris biometrics. *Symmetry* **2018**, *10*, 122. [CrossRef]
18. Li, Y.; Song, L.; He, R.; Tan, T. Anti-Makeup: Learning a bi-level adversarial network for makeup-invariant face verification. *arXiv* **2018**, arXiv:1709.03654.
19. Goodfellow, I.J.; Pouget-Abadie, J.; Mirza, M.; Xu, B.; Warde-Farley, D.; Ozair, S.; Courville, A.; Bengio, Y. Generative Adversial Nets. *arXiv* **2014**, arXiv:1406.2661.

20. Odena, A.; Olah, C.; Shlens, J. Conditional Image Synthesis with Auxiliary Classifier GANs. *arXiv* **2016**, arXiv:1610.09585.
21. Gauthier, J. Conditional Generative Adversarial Nets for Convolutional Face Generation. 2015. Available online: http://cs231n.stanford.edu/reports/2015/pdfs/jgauthie_final_report.pdf (accessed on 15 April 2018).
22. Antipov, G.; Baccouche, M.; Dugelay, J.L. Face aging with conditional generative adversarial networks. *arXiv* **2017**, arXiv:1702.01983.
23. Huang, E.; Zhang, S.; Li, T.; He, R. Beyond face rotation: Global and local perception gan for photorealistic and identity preserving frontal view synthesis. *arXiv* **2017**, arXiv:1704.04086.
24. Li, Z.; Luo, Y. Generate identity-preserving faces by generative adversarial networks. *arXiv* **2017**, arXiv:1706.03227.
25. Zhou, H.; Sun, J.; Yacoob, Y.; Jacobs, D.W. Label Denoising Adversarial Network (LDAN) for Inverse Lighting of Face Images. *arXiv* **2017**, arXiv:1709.01993.
26. Zhang, W.; Shu, Z.; Samaras, D.; Chen, L. Improving heterogeneous face recognition with conditional adversial networks. *arXiv* **2017**, arXiv:1709.02848.
27. Springenberg, J.T. Unsupervised and semi-supervised learning with categorical generative adversarial networks. *arXiv* **2015**, arXiv:1511.06390.
28. Radford, A.; Metz, L.; Chintala, S. Unsupervised representation learning with deep convolutional generative adversarial networks. *arXiv* **2015**, arXiv:1511.06434.
29. Odena, A. Semi-supervised learning with generative adversarial networks. *arXiv* **2016**, arXiv:1606.01583.
30. Salimans, T.; Goodfellow, I.; Zaremba, W.; Cheung, V.; Radford, A.; Chen, X. Improved techniques for training Gans. *arXiv* **2016**, arXiv:1606.03498.
31. Papernot, N.; Abadi, M.; Erlingsson, U.; Goodfellow, I.; Talwar, K. Semi-supervised knowledge transfer for deep learning from private training data. *arXiv* **2016**, arXiv:1610.05755.
32. Fredrickson, B.L. Cultivating positive emotions to optimize health and well-being. *Prev. Treat.* **2003**, *3*. [CrossRef]
33. Fredrickson, B.L.; Levenson, R.W. Positive emotions speed recovery from the cardiovascular sequelae of negative emotions. *Cogn. Emot.* **1998**, *12*, 191–220. [CrossRef] [PubMed]
34. Gallo, L.C.; Matthews, K.A. Understanding the association between socioeconomic status and physical health: Do negative emotions play a role? *Psychol. Bull.* **2003**, *129*, 10–51. [CrossRef] [PubMed]
35. Todaro, J.F.; Shen, B.J.; Niura, R.; Sprio, A.; Ward, K.D. Effect of negative emotions on frequency of coronary heart disease (The Normative Aging Study). *Am. J. Cardiol.* **2003**, *92*, 901–906. [CrossRef]
36. Huang, Y.; Khan, S.M. DyadGAN: Generating facial expressions in dyadic interactions. In Proceedings of the IEEE Conference on Computer Vision and Pattern Recognition Workshops, Honolulu, HI, USA, 21–26 July 2017. [CrossRef]
37. Zhou, Y.; Shi, B.E. Photorealistic facial expression synthesis by the conditional difference adversarial autoencoder. *arXiv* **2017**, arXiv:1708.09126.
38. Lu, Y.; Tai, Y.W.; Tang, C.K. Conditional CycleGAN for attribute guided face image generation. *arXiv* **2017**, arXiv:1705.09966.
39. Ding, H.; Sricharan, K.; Chellappa, R. ExprGAN: Facial expression editing with controllable expression intensity. *arXiv* **2017**, arXiv:1709.03842.
40. Xu, R.; Zhou, Z.; Zhang, W.; Yu, Y. Face transfer with generative adversarial network. *arXiv* **2017**, arXiv:1710.06090.
41. Nojavanasghari, B.; Huang, Y.; Khan, S.M. Interactive generative adversarial networks for facial expression generation in dyadic interactions. *arXiv* **2018**, arXiv:1801.09092.
42. Tian, Y.L.; Kanage, T.; Cohn, J. Robust Lip Tracking by Combining Shape, Color and Motion. In Proceedings of the 4th Asian Conference on Computer Vision, Taipei, Taiwan, 8–11 January 2000.
43. Agarwal, M.; Krohn-Grimberghe, A.; Vyas, R. Facial key points detection using deep convolutional neural network—Naimishnet. *arXiv* **2017**, arXiv:1710.00977.
44. Kazemi, V.; Sullivan, J. One millisecond face alignment with an ensemble of regression trees. In Proceedings of the 2014 IEEE Conference on Computer Vision and Pattern Recognition, Columbus, OH, USA, 23–28 June 2014. [CrossRef]

45. Suh, K.H.; Kim, Y.; Lee, E.C. Facial feature movements caused by various emotions: Differences according to sex. *Symmetry* **2016**, *8*, 86. [CrossRef]
46. Dachapally, P.R. Facial emotion detection using convolutional neural networks and representational autoencoder units. *arXiv* **2017**, arXiv:1706.01509.
47. Lyons, M.J.; Kamachi, M.; Gyoba, J. Japanese Female Facial Expressions (JAFFE). *Database of Digital Images*. 1997. Available online: http://www.kasrl.org/jaffe.html (accessed on 15 April 2018).
48. Huang, G.B.; Ramesh, M.; Berg, T.; Learned-Miller, E. *Labeled Faces in the Wild: A Database for Studying Face Recognition in Unconstrained Environments*; Workshop on Faces in 'RealLife' Images: Detection, Alignment, and Recognition: Marseille, France, 2008.
49. Zhu, X.; Liu, Y.; Qin, Z.; Li, J. Data augmentation in emotion classification using generative adversarial networks. *arXiv* **2017**, arXiv:1711.00648.
50. Facial Expression Recognition (FER2013) Dataset. Available online: https://www.kaggle.com/c/challenges-in-representation-learning-facial-expression-recognition-challenge/data (accessed on 19 October 2017).
51. Lee, K.W.; Hong, H.G.; Park, K.R. Fuzzy system-based fear estimation based on the symmetrical characteristics of face and facial feature points. *Symmetry* **2017**, *9*, 102. [CrossRef]
52. Al-Shabi, M.; Cheah, W.P.; Connie, T. Facial expression recognition using a hybrid CNN-SIFT aggregator. *arXiv* **2016**, arXiv:1608.02833.
53. Lucey, P.; Cohn, J.F.; Kanade, T.; Saragih, J.; Ambadar, Z.; Matthews, I. The Extended Cohn-Kanade Dataset (CK^+); A Complete Dataset for Action Unit and Emotion-Specified Expression. In Proceedings of the 2010 IEEE Computer Society Conference on Computer Vision and Pattern Recognition Workshops, San Francisco, CA, USA, 13–18 June 2010.
54. Dhal, A.; Goecke, R.; Luvey, S.; Gedeon, T. Static facial expressions in tough; data, evaluation protocol and benchmark. In Proceedings of the IEEE International Conference on Computer Vision ICCV2011, Barcelona, Spain, 6–13 November 2011.
55. Mishra, S.; Prasada, G.R.B.; Kumar, R.K.; Sanyal, G. Emotion Recognition through facila gestures—A deep learning approach. In Proceedings of the Fifth International Conference on Mining Intelligence and Knowledge Exploration (MIKE), Hyderabad, India, 13–15 December 2017; pp. 11–21.
56. Quinn, M.A.; Sivesind, G.; Reis, G. Real-Time Emotion Recognition from Facial Expressions. 2017. Available online: http://cs229.stanford.edu/proj2017/final-reports/5243420.pdf (accessed on 15 April 2018).
57. Plutschik, R. The nature of emotions. *Am. Sci.* **2001**, *89*, 344. [CrossRef]
58. Chen, X.; Duan, Y.; Houthooft, R.; Schulman, J.; Sutskever, I.; Abbeel, P. InfoGAN: Interpretable Representation Learning by Information Maximizing Generative Adversarial Nets. *arXiv* **2016**, arXiv:1606.03657.
59. Dlib Library. Available online: http://blog.dlib.net/2014/08/real-time-face-pose-estimation.html (accessed on 12 November 2017).

Article

Accurate Age Estimation Using Multi-Task Siamese Network-Based Deep Metric Learning for Frontal Face Images

Yoosoo Jeong, Seungmin Lee, Daejin Park * and Kil Houm Park *

School of Electronics Engineering, Kyungpook National University, Daegu 41566, Korea; ysjung@ee.knu.ac.kr (Y.J.); lsm1106@knu.ac.kr (S.L.)
* Correspondence: boltanut@knu.ac.kr (D.P.); khpark@ee.knu.ac.kr (K.H.P.); Tel.: +82-53-950-5548 (D.P.)

Received: 26 July 2018; Accepted: 4 September 2018; Published: 6 September 2018

Abstract: Recently, there have been many studies on the automatic extraction of facial information using machine learning. Age estimation from frontal face images is becoming important, with various applications. Our proposed work is based on a binary classifier that only determines whether two input images are clustered in a similar class and trains a convolutional neural network (CNN) model using the deep metric learning method based on the Siamese network. To converge the results of the training Siamese network, two classes, for which age differences are below a certain level of distance, are considered as the same class, so the ratio of positive database images is increased. The deep metric learning method trains the CNN model to measure similarity based only on age data, but we found that the accumulated gender data can also be used to compare ages. Thus, we adopted a multi-task learning approach to consider the gender data for more accurate age estimation. In the experiment, we evaluated our approach using MORPH and MegaAge-Asian datasets, and compared gender classification accuracy only using age data from the training images. In addition, using gender classification, our proposed architecture, which is trained with only age data, performs age comparison using the self-generated gender feature. The accuracy enhancement by multi-task learning, i.e. simultaneously considering age and gender data, is discussed. Our approach results in the best accuracy among the methods based on deep metric learning on MORPH dataset. Additionally, our method has better results than the state of the art in terms of age estimation on MegaAge-Asian and MORPH datasets.

Keywords: convolutional neural network (CNN); deep metric learning; multi-task learning; image classification; age estimation

---

1. Introduction

Machine learning-based age estimation from face images is becoming more and more important because it is widely used for individual authentication [1], forensic research [2], security control [3], human–computer interaction [3] and social media [4]. Recently, there have been many studies using deep learning based on CNNs [3], such as AlexNet [5], VggNet [6], and Inception [7], with wide use for image classification and image detection. CNN-based learning, as one of the machine learning-based approaches, enables automatic and accurate feature extraction and classification for sample sets that are too large for humans to describe all cases of matching patterns. AlexNet, VggNet, and Inception have recently been used for multi-class classification, and they are widely used as the base models of CNN.

Deep expectation (DEX) [4] is an age estimation approach based on CNN models. It uses VggNet to resolve multi-class classification problems for age estimation and adopts a method to estimate the appropriate age through expectation value calculation, for which the trained results in the softmax layer are considered the probability in the corresponding class. Instead of considering the age estimation problem from the perspective of multi-class classification, this approach applies multi-task CNN by considering the age classification problem as a regression-based problem by estimating continuous variables [8].

As another approach, a binary classifier with shallow layers is applied for all classes of age instead of using a CNN model with deep layers. The final age estimation is deducted through the ranking-based comprehensive combination of all results by each binary classifier [9]. This ranking CNN is one of the existing machine learning methods using the cascaded-based combination of the results of binary classifiers.

1.1. Motivation

The above approaches aim to estimate absolute age from the input face images directly, but it is not easy to estimate absolute age accurately without any reference data [10]. To overcome this limitation, Abousaleh et al. [11] introduced a new approach, called comparative region convolutional neural network (CRCNN). Input face images are compared with reference images to determine whether they are older or younger for age estimation. Our study was also inspired by this CRCNN, comparing the age relatively instead of directly estimating absolute age, so we adopted the deep metric learning method to train the logic of comparing age in the CNN model. Deep metric learning reduces the complex classification task to the nearest neighbor problem [10]. In addition, this approach has the advantage that it makes use of relationships using more data.

A Siamese network [12] is widely used as a deep metric learning-based approach. Two input images are applied to two CNN models, and then each input image is mapped to a point in multi-dimensional space, where the similarity of the two input images is described as the corresponding distance. These CNN models are trained using the loss function, by which the points are closely clustered in the case of higher similarity. A well-trained Siamese network generates well-clustered data for the training images. The input image can be accurately labeled by selecting the nearest clustered data compared to the features extracted from input images. Here, the nearest neighbor selection process corresponds to our approach of estimating the labels by comparing the input images with the training images.

However, Siamese network-based deep metric learning has the drawback of difficulty in converging the results. When this learning method is applied for age estimation, all remaining classes except the correct class are negative so divergence often occurs in the learning process. Related to this issue, CRCNN trains a Siamese network using loss function to determine whether the age is younger using two images instead of comparing the similarities. Additionally, CRCNN proposes a selection approach for specific images compared with the input images. This avoids the side effect of continuously learning with negative reference images.

1.2. Contribution

With these motivations, by applying a Siamese network-based deep metric learning for exact age estimation, we propose a method to converge the process of Siamese network learning. Our proposed approach allows a certain level of error tolerance to increase the ratio of positive data, so that it can perform comparisons for all images in the database, while decreasing the possibility of divergence in the training process.

Additionally, the deep metric learning method trains the CNN model to measure similarity based only on age data, but we found that the accumulated gender data can also be used to compare the age. Thus, we adopted a multi-task learning approach to consider the gender data for more accurate age estimation. Multi-task learning is a method to train CNN models simultaneously with multiple tasks to effectively assist in the training. This method enables the CNN models to be trained to simultaneously perform the age estimation tasks and separate tasks to classify the gender, so that more relationship data can be involved, which is helpful to increase the performance in terms of accuracy.

The whole process is as follows. We use Inception V3 for CNN model [13], which is pre-trained with ImageNet [14], and perform the feature-embedding by considering the value of the fully connected layer. The loss function is designed to train our architecture to decrease the distance between feature vectors when two images in batch are in the same class, as well as to increase the distance between feature vectors in the case of differences in class for two images. In this step, we allow a certain level of error tolerance for determining whether two images are in the same class. We define the two feature vectors for measuring age similarity and for measuring gender similarity, respectively. Two feature vectors are simultaneously trained to perform the multi-task learning method.

After training step with these conditions, the feature vectors for all training databases are extracted and the distribution of the clustered data with respect to age similarity can be obtained.

In the test step, the featured vector of an input image is selected with the nearest one in the feature space to compare the relative location in the clustered data distribution.

This paper is organized as follows. Section 2 explains in detail our architecture to perform the learning for age estimation. Section 3 shows the experimental results using the proposed approach, and discusses the performance of the proposed models. Section 4 provides the conclusion of this study.

2. Proposed Architecture

The structure of the neural network in our proposed architecture, which is a Siamese network, is described in Figure 1 [12]. As shown in Figure 1, the structure and weights in these two networks are completely equivalent. The outputs of two CNN models for input Images A and B are used in loss function and the relationship is determined according to the design of loss function. These two networks are used to apply the loss function for the inference as a result of two input Images A and B.

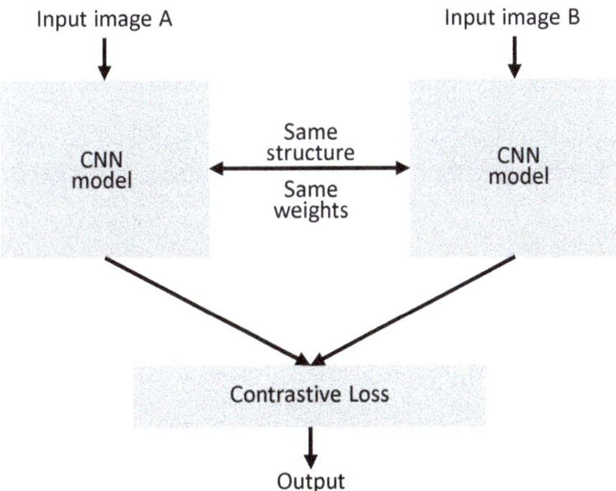

Figure 1. Structure of the Siamese network.

In this paper, instead of using two Siamese network-based CNN models for age comparison from two input images, we apply the contrastive loss function using the inference results for the corresponding images by selecting two images from a batch of training models in a single network.

Figure 2 shows an illustration of the overall algorithm. Inception V3 is used for the construction of the CNN model, but with a fully connected layer, not using a softmax layer. To apply the multi-task learning to estimate age and gender simultaneously, one more fully connected layer is constructed. The first fully connected layer performs age comparison and the second fully connected layer assists age comparison by performing the gender comparison task.

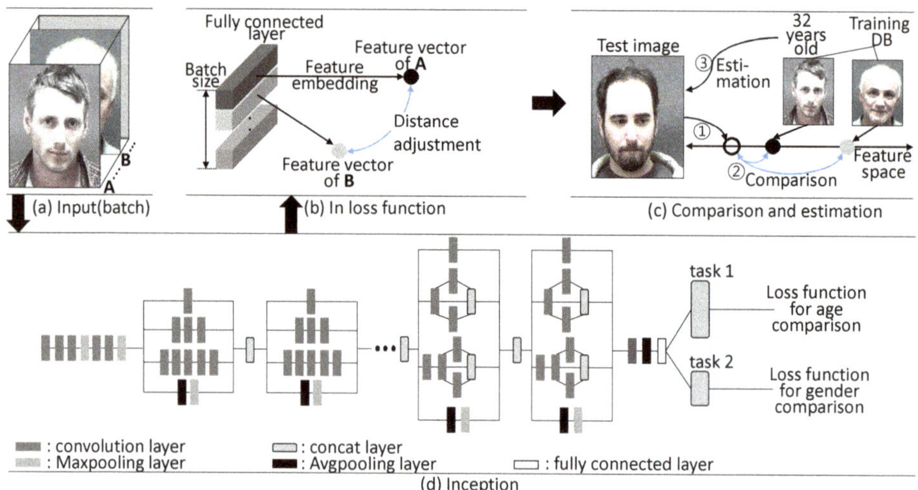

Figure 2. Illustration of the proposed algorithm.

As shown in Figure 2a, two input images are selected from the batch, by considering all selectable combinations. The selected A and B images are mapped to the feature vector which is a final output of the fully connected layers in Figure 2d. In the proposed loss function, the gradient value is propagated into the network to decrease the distance between feature vectors when two images in batch are in same class, as well as to increase the distance between feature vectors then the two images are in different classes, as shown in Figure 2b. Our architecture is trained using the proposed algorithm to determine the similarity between two input images.

In the test step, feature vector of test image is compared to feature vectors of the entire training database to perform age estimation by selecting the most similar age class, as shown in Figure 2c. The detailed process of the proposed algorithm is as follows.

2.1. Inception V3

The proposed algorithm in this paper adopted Inception V3 [13], which is an enhanced version with batch normalization and filter size reduction.

Figure 3 compares module of the Inception model and module of the Inception V3 model. In the Inception model, the filter sizes are 5×5 and 1×1, but the Inception V3 model uses 1×1 and $N \times 1$ filters continuously; as a result, the calculation cost and the number of parameter coefficients are reduced. In this paper, we adopt the Inception V3 model and configure the $(N = 3) \times 1$ filter. To perform Siamese network-based deep metric learning using this Inception V3 model, the final output of fully connected layers is used as the feature vector instead of using the softmax layer, as shown in Figure 2b.

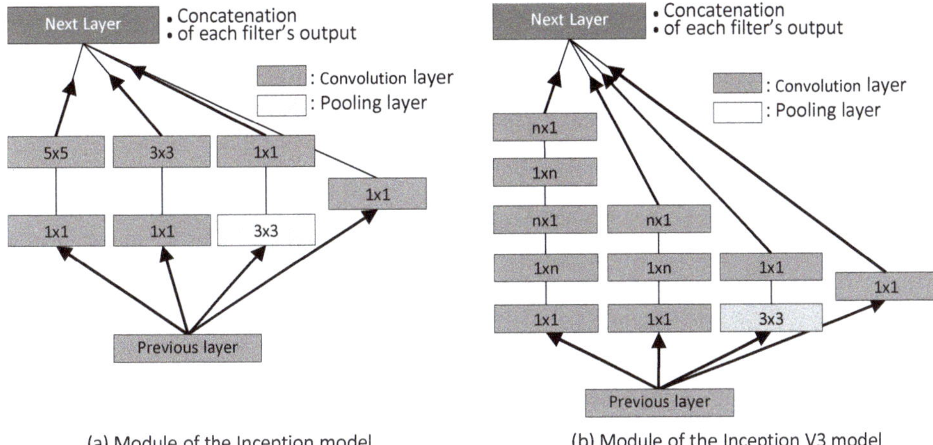

Figure 3. Inception module.

2.2. Selection of Two Images and the Feature-Embedding Process

To implement the Siamese network using a single network, two images are selected from the batch, as shown in Figure 4, and they are used to measure the similarity. The comparison repeats the number of available combinations by selecting two images from the batch. Unlike in the previous CRCNN, this approach performs the comparisons and trains the model between all images in the batch instead of selecting only specific images [11].

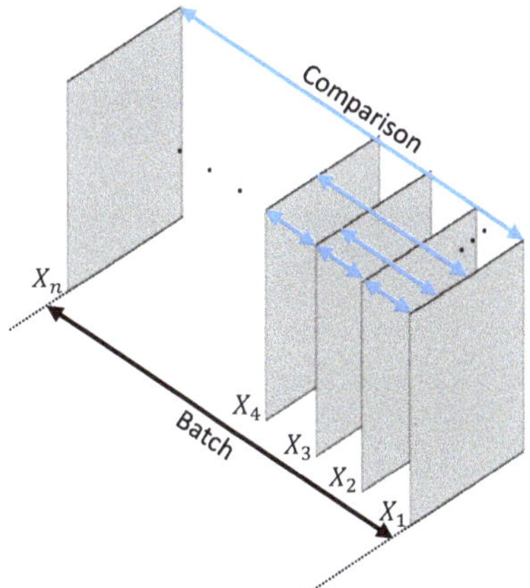

Figure 4. Image selection for comparison in batch.

The two selected images X_i, X_j in the batch, as shown in Figure 5, are mapped and shrunk to the final fully connected layer in N_a dimensions, which is described using Inception V3. The shrunk data are represented with the corresponding features $FV(X_i), FV(X_j)$, in which integers i and j are indices in the batch.

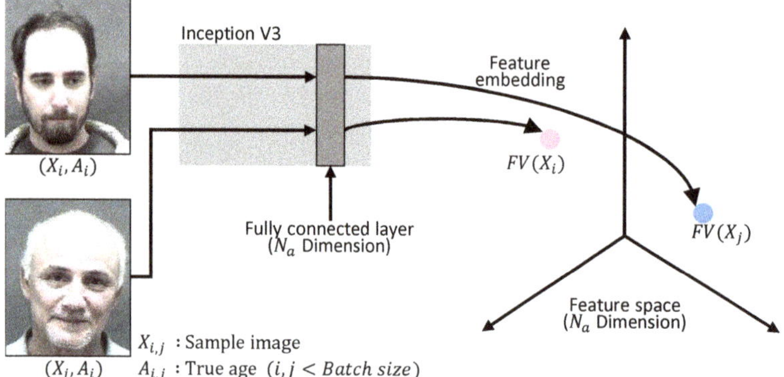

Figure 5. Feature-embedding.

2.3. Distance as Similarity between Two Images

The feature vectors are extracted by the Inception V3-based feature-embedding method, as shown in Figure 5.

The proposed algorithm aims to effectively train the model by mapping the feature vectors into feature space so that similar images are clustered with smaller distance. Therefore, similarity between two images and distance between feature vectors of two images have a reciprocal proportion relationship. The distance between the feature vectors is calculated using $L^1 - norm$, which calculates the absolute distance of the corresponding value in each dimension with the following equation in terms of the distance D between two feature vectors.

$$D = ||FV(X_i) - FV(X_j)||_{L^1} \tag{1}$$

Some previous approaches [11,13] use the Euclidean distance calculation method called norm2, but the preferred approach in previous studies has been to use norm1 instead of norm2 for Siamese networks [12].

In this paper, we define the distance using $L^1 - norm$ and we can successfully converge the training result, as evaluated in the experiment.

2.4. Loss Function for the Training Comparison Task

Feature vector comparison, as a representative descriptor for a given image, is equivalent to comparing the image itself. Our proposed approach defines the loss function and trains the comparison task of the CNN model so that the extracted features are relatively positioned in the feature space in terms of the similarity of two feature vectors.

The loss function used in this paper is described as follows. The loss function corresponds to the contractive loss function in the Siamese network, which is introduced as a contrastive loss function [12].

$$loss = (1 - \overline{Z})L^-(D) + (\overline{Z})L^+(D) \tag{2}$$

$\overline{Z}$ is a Boolean function that outputs 1 in the case of two similar images; otherwise, it outputs 0. L^- has to satisfy the condition in the manner of a decreasing function, and L^+ of an increasing function, as shown in the following equation.

$$\overline{Z} = \begin{cases} 1, & \text{if two images are considered as same class} \\ 0, & \text{otherwise} \end{cases} \tag{3}$$

$$L^{-}(x) = 2 \times Qe^{-\frac{2.77}{Q}x}, L^{+}(x) = \frac{2}{Q} \times x^{2} \tag{4}$$

Q is a constant to determine the upper limit of dissimilarity, which is 100 in this paper. Figure 6 is a graph to describe the loss function in terms of the distance between feature vectors. $\overline{Z}$ is 1 in the case of two similar images in the same class, and the L^{+} term remains. The gradient is propagated into the network so that the distance is reduced to minimize the loss in the designed loss function. $\overline{Z}$ is 0 in the case of two images that are considered to be in different classes, and the L^{-} term remains. The gradient is propagated into the network so that the distance is increased for the decreased loss function. With these operations in the network, the weights for feature vector extraction is updated.

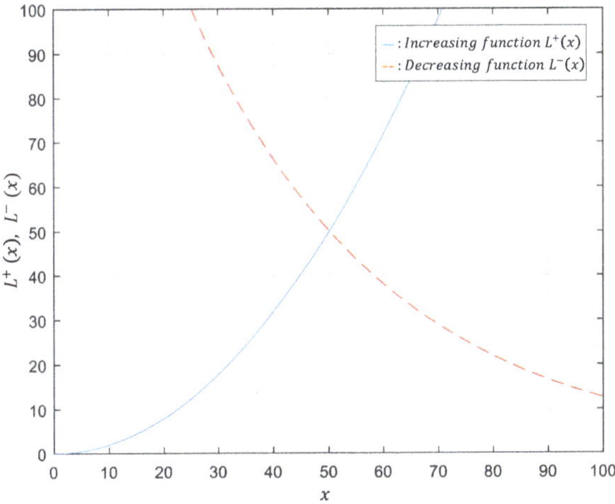

Figure 6. A designed loss function for the proposed algorithm.

Because this designed loss function is used to train the network to determine the distance between feature vectors, there is no inefficiency limiting the basis of the mapping plane. However, unlike the trained database, the proposed method has to search and determine a nearest neighbor from feature vectors. In addition, an approach using this loss function enables the multi-class classification for age estimation of various bands to be simplified as a binary classification problem which only measures the similarity. It mitigates imbalance of the accuracy over all classes, which is caused by the biased training database. However, if this loss function is applied to the binary classifier as it is, the images in the same age class are considered positive, and all other classes are negative; as a result, the trained database becomes imbalanced due to the large number of classes, which is why Siamese networks do not easily converge the training results.

To resolve this issue, CRCNN adopts a technique to select the comparison images in advance to prevent the network from being continuously trained with the negative database. Instead of comparing the similarities in age, it redesigns the loss function to only determine whether the age is younger or older; as a result, it can converge the training results of the Siamese network.

Our approach succeeds in converging the training result by adopting a method to increase the ratio of the positive data, for which the Boolean function $\overline{Z}$ determining age class allows for error tolerance. For example, if three years is allowed as a margin, the loss function considers classes between $N-3$ and $N+3$ years old to be the same class. The proposed technique is helpful to increase the ratio of positive data, so the entire process of training the CNN model is not negatively influenced by the error tolerance.

In fact, while our approach loses discrimination by class in the CNN model with the margin-allowed error, it results in more accurate age estimation by enabling all comparisons for all age ranges. Even though a specific feature vector is involved with the class within a certain range of marginal error tolerance, clustering can be processed further with accuracy of the margin value, by comparing with the feature vector within (margin+1) and −(margin+1) compared to the currently clustered age. The entire clustering procedure using the proposed approach is described in Figure 7.

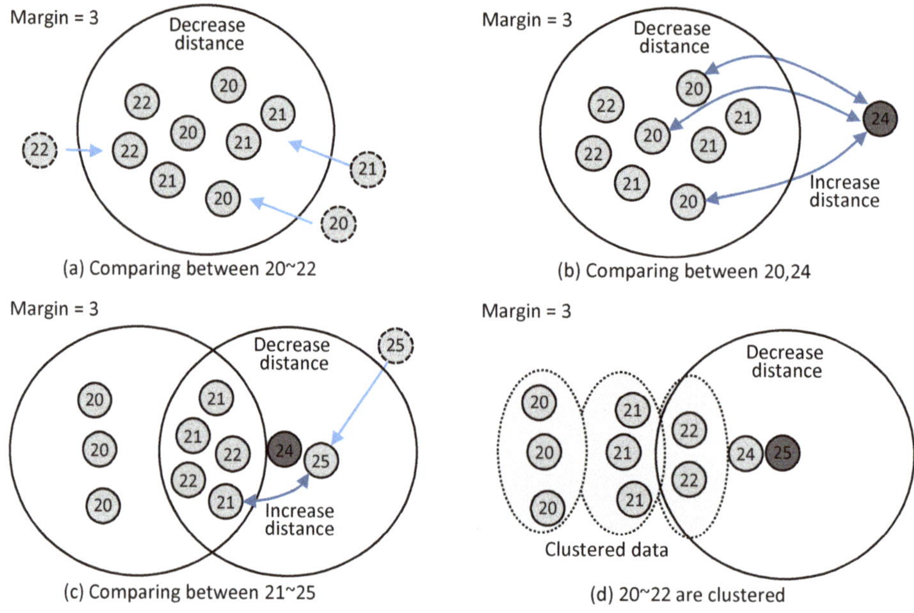

Figure 7. Clustering process allowing marginal error tolerance.

Figure 7 assumes that the margin is defined as 3; the feature vectors of the images are compared and clustered using the proposed loss function. For example, as shown in Figure 7a, if only the feature vectors of the images that are 20–22 years old are compared, then all images are considered similar because the margin is defined as 3, so only the distance decreases, but the clustering does not proceed further. This means that the estimation accuracy is three years. However, as shown in Figure 7b, if the feature vector for an image classified as 24 years old is compared to one classified as 20 years old, the network is trained to increase the distance, so the feature vector of age 24 is clustered to be positioned far away. As shown in Figure 7c, the network is trained so that the feature vector for 21–22 age is clustered to be closely positioned, because ages 20–21 and 24 are within the margin, which can be considered the same class. When a feature vector with 25 is compared, 22 and 25 are considered the same class through the same process, so the network is trained to have a close distance between 22 and 25. As a result, the feature vectors of 20, 21, 22, 24, and 25 are separately clustered, so we can distinguish the age of the images with an accuracy of one year.

2.5. Age Estimation

In the test step using the database trained by the proposed approach, the age estimation process initially involves calculating the feature vectors in N_a dimensions to search for similar images compared to the trained database. Because the CNN model has already been trained to determine the age similarity, the test model comparing the input image is prepared with the clustered feature vectors. The feature vector for the input image is extracted using the same CNN model, and then compared with the clustered data in the test model. The test process involves age estimation performed by calculating mean age of among M_{th} nearest neighborhoods. The distance-based nearest neighborhood search method is also based on $L^1 - norm$ which is used in the training process. The entire test process is described in Figure 8.

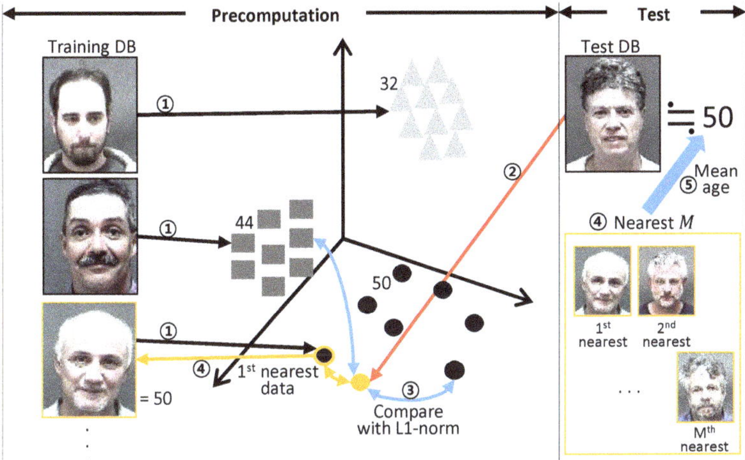

Figure 8. Age estimation process selecting the nearest neighborhood in the feature space.

2.6. Multi-Task Learning for Age and Gender Estimation

The loss function for the proposed method is designed to train the CNN model with age similarity as the relation of classes. Even though the CNN model is trained to determine a similar level using the age data, it can be further trained by clustering the classes closely with similar images using detailed conditions, such as face angle, hair length, and beard. An algorithm that determines age using various conditions, in addition to the absolute age data, is more appropriate. That is why the detailed conditions are automatically configured and applied to the training model by only defining the age-based similarity.

With this concept, we first tried gender classification using the model trained with only the age data, and then we measured the accuracy of the gender-matching result. We found that our approach using only age-based estimation could classify the gender with 81.23% accuracy compared to the result of gender-based classification. The result is summarized in Table 2. The result gave us the following two insights. First, our approach internally uses gender-based conditions to perform the age estimation. Second, the gender data can be an important clue to estimate age. In fact, the 81.23% accuracy of gender classification based on age data means that the age estimation is tightly coupled with gender.

Based on this speculation, our approach adopts the multi-task learning approach so that it additionally provides gender data to the trained model when comparing age. The multi-task learning simultaneously trains the model to increase the performance in terms of accuracy of age estimation. If the individual tasks have a cross-coupled relationship, the multi-task learning approach enables the model to be trained by selecting commonly important variables in the multiple tasks. Utilizing the ability to train the model considering the relationships between tasks, we could assist the age estimation with gender data, thus training the model to consider age and gender simultaneously.

The multi-task learning technique applied in this paper is described in Figure 9. A fully connected layer in N_g dimensions is added for the gender comparison used to compare age in Inception V3. We also designed a loss function to train the logic of the gender comparison so that the weights in the layer are updated in a similar way as in the age comparison algorithm. The margin of comparison in the loss function is 0, and it divides the positive and negative data on the basis of gender. This additional task for gender comparison is temporarily used to assist the data in training the age estimation logic. The finally calculated loss is the sum of the loss by the age estimation and gender classification.

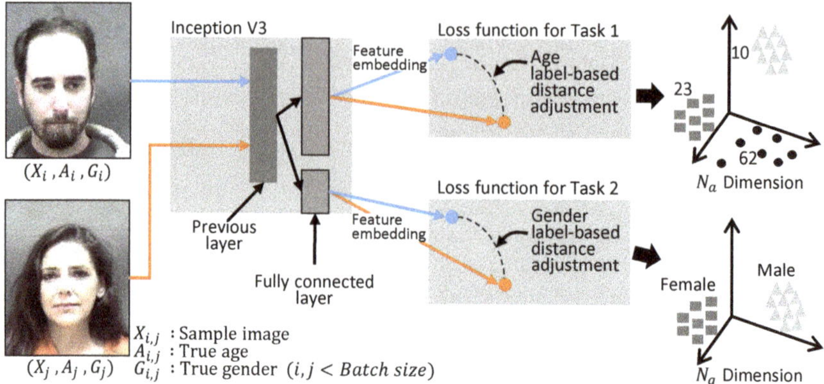

Figure 9. Multi-task learning for the age algorithm considering age and gender simultaneously.

3. Experimental Results and Discussion

The purpose of this experiment was to verify the age estimation accuracy performance based on our architecture in an open image database. We implemented our algorithm using TensorFlow [15], an open source deep learning framework based on Python. We used Inception V3 for CNN model [13], which was pre-trained with ImageNet [14]. The batch size was 128, the image size was 227 and dropout was performed with a probability of 50%. The first fully connected layer's dimension, N_a, tasked with measuring age similarity, was 70, and N_g, i.e. the dimension of the second fully-connected layer for measuring gender similarity, was 10. Each dimension was experimentally selected. In the gradient descent procedure to optimize network weights, the Adadelta [16] method was used. The margin-allowed error, newly defined in our proposed method, was set to 4. This means that, if the difference between age was less than 4, the two ages were considered to be in the same class. In the test step, mean age of the nearest 20 ($M = 20$) was calculated for prediction. The age estimation performance was evaluated by mean absolute error (MAE), which is generally used in previous research as defined in the following equation. MAE indicates how close a prediction is to the true age.

$$MAE = \frac{\sum_{i=1}^{n} |A_i - \tilde{A}_i|}{n} \quad (5)$$

$\tilde{A}_i$ and A_i are the estimate and true age of the sample image j, and n is the total number of samples. We also calculated the cumulative score (CS) [17–19]. CS indicates the percentage of samples correctly estimated in the range of $[A_i - T, A_i + T]$, a neighbor range of the true age where T is the parameter representing the tolerance. CS is calculated using the following equation.

$$CS(T) = 100 \times \frac{\sum_{i=1}^{n} [|A_i - \tilde{A}_i| <= T]}{n} \quad (6)$$

Here, [.] is the truth-test operator. A higher value of $CS(T)$ means a better performance of the architecture. We experimented with two public datasets. The first was the MORPH database [20]. There are 55,132 face images from more than 13,000 subjects in this database. The ages of the face images range from 16 to 77. The frontal face images are from different races, among which African faces account for about 77%, European faces account for about 19% and the remaining 4% include Hispanic, Asian, Indian, and other races [11]. The second was MegaAge-Asian [21]. It contains 40,000 face images of Asians with ages from 0 to 70. Table 1 shows the size of each dataset and the corresponding splits for training and testing. We first selected test images randomly and the remaining images were used as training images. Therefore, there is no intersection between training and test sets.

Table 1. The proposed method was evaluated using two datasets.

DB Name	The Number of Training Images	The Number of Test Images
MegaAge-Asian	40,000	4000
MORPH	45,132	10,000

The proposed architecture was trained with each dataset in Table 1. We experimented to verify the performance of our method, as described in the following sections.

3.1. Toy Example: Visualization of Feature Embedding Computed by Our Method Using a Subset of the MORPH Dataset

To verify that the clustering process improves the accuracy of the margin value, feature vectors were visualized using a small subset of the MORPH dataset. For visualization on two-dimensional space and to facilitate convergence, we collected face images with ages from 16 to 63 (only 48 classes) and each class had 1–3 images randomly. Hyper parameters for the toy example are as follows. The batch size was 48, and the dimension of the feature vector was 2 for visualization on two-dimensional space. In the case of the toy example, the margin value was set as 2. After the training step with these conditions, extracted feature vectors were clustered, as shown in Figure 10. The vertical axis is the true age of each feature vector and the others are axes of feature space. Most of the feature vectors were well-clustered, as shown in the zoomed graph (red box). The clustering process had an accuracy of one year but our CNN model had an accuracy of two years, thus putting those images that were two years younger or two years older in the same class.

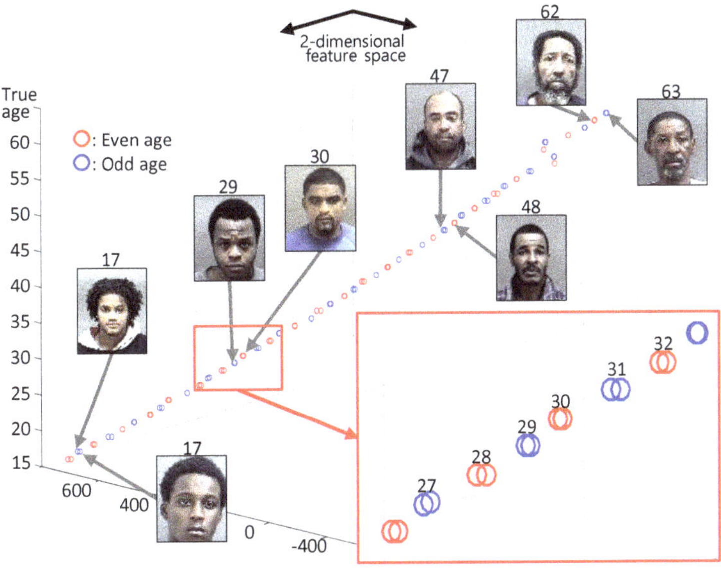

Figure 10. Visualization of feature embedding with the toy example.

3.2. Multi-Task Learning for Age and Gender Estimation

The first row of Table 2 is the result of the gender classification rate on the MORPH dataset using only age data. Even though gender data were not used, the gender classification rate was quite high. The classification rate was much lower than in the other CNN model using gender data. Even AlexNet, which is a relatively simple model, had a better classification rate. However, 80% accuracy means that the age estimation is tightly coupled with the gender. Therefore, we tried using gender data in the CNN model by applying multi-task learning for age and gender estimation simultaneously. The results of the experiment before and after applying the multi-task learning method are shown in Figure 11. The MAE of our method with multi-task learning slightly decreased from 2.24 to 2.28, but $CS(T)$ values were improved. In particular, the $CS(5)$ value increased by about 2%. Therefore, performance was improved by using gender data to estimate age through multi-task learning.

Table 2. Gender classification rates on the MORPH dataset.

Method	Accuracy (%)
Our method without gender data	81.23
Alexnet [5] with gender data	97.38
Inception V3 [7] with gender data	99.1

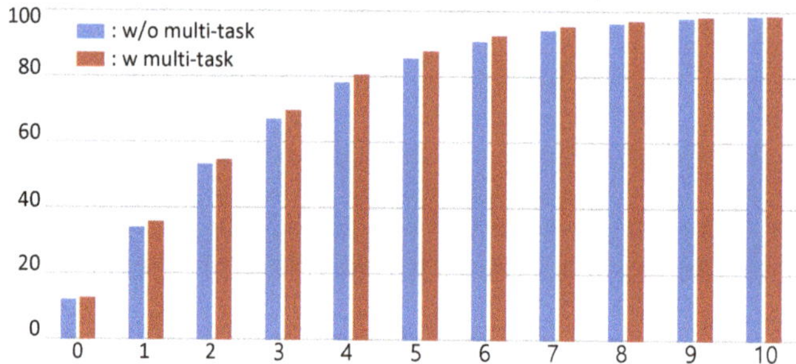

Figure 11. Comparison of our method with and without multi-task learning.

3.3. Comparison with Deep Metric Learning-Based Approaches on the MORPH Dataset

Table 3 shows the age estimation result of the dataset and a comparison with traditional methods based on deep metric learning. The MAE of our method was 2.24, indicating better accuracy than the MAE of the CRCNN [11], which is 3.74. This means that applying our method of deep metric learning based on a Siamese network is suitable for age estimation. Moreover, M-LSDML [22], the latest age estimation method based on deep metric learning, has a slightly lower MAE than our method. Additionally, the MAE of the ResNet with each loss function for deep metric learning is shown.

Table 3. Age estimation results on test images of the dataset and a comparison with traditional deep metric learning methods.

Method	Kinds of Loss Function	MORPH(MAE)
Our method	Revised contrastive loss function	2.24
Our method with multi-task learning	Revised contrastive loss function	2.28
CRCNN [11]	Contrastive loss function	3.74
M-LSDML [22]	Custom-defined loss function	2.89
ResNet (contrastive loss) [22]	Contrastive loss function	3.72
ResNet (triplet hinge loss) [22]	Triplet hinge loss function	3.59
ResNet (lifted structural loss) [22]	Lifted structural loss function	3.24

3.4. Comparison with State-of-Art Method on Each Dataset

In addition, we compared the state-of-the-art methods. Most techniques using the MegaAge-Asian dataset evaluate age estimation performance by $CS(T)$, as shown in Table 4. Our method achieved a slightly higher score than the other methods on the MegaAge-Asian dataset. In the case of techniques using the MORPH dataset, the MAE is widely used to evaluate the age estimation performance. In Table 5, the MAE of each technique is shown. In the experiment on MORPH dataset, our method achieved the best MAE (2.24).

Table 4. Comparison of $CS(T)$ with state-of-the-art methods on the MegaAge-Asian dataset (* face alignment method is applied, ** additional labels are used).

Method	$CS(3)$	$CS(5)$
Our method	69.70	84.64
MobileNet [23]	44.0	60.6
DenseNet [24]	51.7	69.4
Zhang et al. [25] **	64.08	82.43
SSR-Net [26] *	54.9	74.1

Table 5. Comparison of MAE with state-of-the-art methods on the MORPH dataset (* face alignment method is applied, ** additional labels are used).

Method	MAE
Our method	2.24
Our method with multi-task **	2.28
Ranking-CNN [9]	2.96
DEX [4] *	3.25
DEX w IMDB [4] *	2.68
Zhang et al. [25] **	2.87
Zhang et al. w IMDB-WIKI [25] **	2.52
SSR-Net [26] *	2.52

In terms of age estimation, the accuracy of our method is improved with respect to CS value and MAE by using more data from relationship between images. However, to deal with bigger datasets, comparing all images may not be an efficient strategy because of the increased computation and clustered data. Our architecture has the disadvantage of longer training time: in the case of applying our multi-task method in MORPH datasets, our architecture needs 275 epochs to converge. In future work, to reduce the training time, we will consider a strategy of automatically selecting images which can be references to compare with training dataset and used for a gallery. This strategy is more appropriate to apply for bigger and more varied datasets (e.g., FG-net and IMDB-WIKI). Additionally, for optimizing our method, more analysis on dimension of feature vector, the consideration of simpler networks with statistical significance according to random initialization and more efficient loss function are needed, which will be researched in future work.

4. Conclusions

This study was motivated by the fact that training a CNN model based on age comparison is easier than directly estimating the absolute age. The proposed approach trains a CNN model for age comparison using a Siamese network-based deep metric learning method. We designed a binary classifier, which is applied to train a Siamese network to cluster the classes within the margin of tolerance as the same class, allowing us to successfully train the Siamese network by adopting $L^1 - norm$ instead of $L^2 - norm$. The experimental test indicated that the proposed approach itself performs the gender classification in processing the age estimation, thus we tried training the CNN model by comparing age and gender simultaneously using the multi-task learning technique. The proposed method was evaluated using the MORPH dataset. Although our architecture needs many epochs to converge, it results in better performance. There was also an additional enhancement using multi-task learning for age and gender compared to CRCNN, the original Siamese network-based deep metric learning, and the latest M-LSDML. Additionally, our method also has better results than the state of the art on MegaAge-Asian and MORPH datasets. In future work, more analysis is needed to reduce the training time by selecting reference images to compare rather than comparing all images.

Author Contributions: Y.J. designed the entire core architecture and performed the hardware/software implementation and experiments; S.L. validated the experimental results by the proposed framework; and K.H.P. proposed the key concept and algorithm of the proposed architecture. D.P. is the corresponding author.

Funding: This study was supported by the BK21 Plus project funded by the Ministry of Education, Korea (21A20131600011).

Conflicts of Interest: The authors declare no conflict of interest. The founding sponsors had no role in the design of the study; in the collection, analyses, or interpretation of data; in the writing of the manuscript, and in the decision to publish the results.

Abbreviations

The following abbreviations are used in this manuscript:

CNN	Convolutional Neural Network
DEX	Deep EXpectation
CRCNN	Comparative Region Convolution Neural Network
MAE	Mean Absolute Error
CS	Cumulative Score

References

1. Ling, H.; Soatto, S.; Ramanathan, N.; Jacobs, D.W. A Study of Face Recognition as People Age. In Proceedings of the 2007 IEEE 11th International Conference on Computer Vision, Rio de Janeiro, Brazil, 14–21 October 2007; pp. 1–8. [CrossRef]
2. Alkass, K.; Buchholz, B.A.; Ohtani, S.; Yamamoto, T.; Druid, H.; Spalding, K.L. Age estimation in forensic sciences: Application of combined aspartic acid racemization and radiocarbon analysis. *Mol. Cell. Proteom. (MCP)* **2010**, *95*, 1022–1030. [CrossRef] [PubMed]
3. Han, H.; Otto, C.; Jain, A.K. Age estimation from face images: Human vs. machine performance. In Proceedings of the 2013 International Conference on Biometrics (ICB), Madrid, Spain, 4–7 June 2013; pp. 1–8.
4. Rothe, R.; Timofte, R.; Van Gool, L. Deep Expectation of Real and Apparent Age from a Single Image without Facial Landmarks. *Int. J. Comput. Vis.* **2018**, *126*, 144–157. [CrossRef]
5. Krizhevsky, A.; Sutskever, I.; Hinton, G.E. ImageNet Classification with Deep Convolutional Neural Networks. In *Advances in Neural Information Processing Systems 25*; Pereira, F., Burges, C.J.C., Bottou, L., Weinberger, K.Q., Eds.; Curran Associates, Inc.: Red Hook, NY, USA, 2012; pp. 1097–1105.
6. Simonyan, K.; Zisserman, A. Very Deep Convolutional Networks for Large-Scale Image Recognition. *arXiv* **2014**, arXiv:1409.1556. [1409.1556].
7. Szegedy, C.; Liu, W.; Jia, Y.; Sermanet, P.; Reed, S.E.; Anguelov, D.; Erhan, D.; Vanhoucke, V.; Rabinovich, A. Going Deeper with Convolutions. *arXiv* **2015**, arXiv:1409.4842. [1409.4842].
8. Yin, X.; Liu, X. Multi-Task Convolutional Neural Network for Face Recognition. *arXiv* **2017**, arXiv:1702.04710. [1702.04710].
9. Chen, S.; Zhang, C.; Dong, M.; Le, J.; Rao, M. Using Ranking-CNN for Age Estimation. In Proceedings of the 2017 IEEE Conference on Computer Vision and Pattern Recognition (CVPR), Honolulu, HI, USA, 21–26 July 2017; pp. 742–751.
10. Song, H.O.; Xiang, Y.; Jegelka, S.; Savarese, S. Deep Metric Learning via Lifted Structured Feature Embedding. *arXiv* **2015**, arXiv:1511.06452. [1511.06452].
11. Abousaleh, F.S.; Lim, T.; Cheng, W.H.; Yu, N.H.; Hossain, M.A.; Alhamid, M.F. A novel comparative deep learning framework for facial age estimation. *EURASIP J. Image Video Process.* **2016**, *2016*, 47. [CrossRef]
12. Chopra, S.; Hadsell, R.; LeCun, Y. Learning a similarity metric discriminatively, with application to face verification. In Proceedings of the 2005 IEEE Computer Society Conference on Computer Vision and Pattern Recognition (CVPR'05), San Diego, CA, USA, 20–25 June 2005; pp. 539–546.

13. Szegedy, C.; Vanhoucke, V.; Ioffe, S.; Shlens, J.; Wojna, Z. Rethinking the Inception Architecture for Computer Vision. *arXiv* **2015**, arXiv:1512.00567. [1512.00567].
14. Russakovsky, O.; Deng, J.; Su, H.; Krause, J.; Satheesh, S.; Ma, S.; Huang, Z.; Karpathy, A.; Khosla, A.; Bernstein, M.; et al. ImageNet Large Scale Visual Recognition Challenge. *Int. J. Comput. Vis. (IJCV)* **2015**, *115*, 211–252. [CrossRef]
15. Abadi, M.; Barham, P.; Chen, J.; Chen, Z.; Davis, A.; Dean, J.; Devin, M.; Ghemawat, S.; Irving, G.; Isard, M.; et al. TensorFlow: A System for Large-Scale Machine Learning. In Proceedings of the OSDI'16 Proceedings of the 12th USENIX Conference on Operating Systems Design and Implementation, Savannah, GA, USA, 2–4 November 2016; pp. 265–283.
16. Zeiler, M.D. ADADELTA: An Adaptive Learning Rate Method. *arXiv* **2012**, arXiv:1212.5701. [1212.5701].
17. Zhang, Y.; Yeung, D.Y. Multi-task warped Gaussian process for personalized age estimation. In Proceedings of the 2010 IEEE Computer Society Conference on Computer Vision and Pattern Recognition, San Francisco, CA, USA, 13–18 June 2010; pp. 2622–2629.
18. Guo, G.; Mu, G. Simultaneous dimensionality reduction and human age estimation via kernel partial least squares regression. In Proceedings of the CVPR 2011, Colorado Springs, CO, USA, 20–25 June 2011; pp. 657–664.
19. Guo, G.; Fu, Y.; Dyer, C.R.; Huang, T.S. Image-Based Human Age Estimation by Manifold Learning and Locally Adjusted Robust Regression. *IEEE Trans. Image Process.* **2008**, *17*, 1178–1188. [CrossRef] [PubMed]
20. Ricanek, K.; Tesafaye, T. MORPH: A longitudinal image database of normal adult age-progression. In Proceedings of the 7th International Conference on Automatic Face and Gesture Recognition (FGR06), Southampton, UK, 10–12 April 2006; pp. 341–345.
21. Kemelmacher-Shlizerman, I.; Seitz, S.M.; Miller, D.; Brossard, E. The MegaFace Benchmark: 1 Million Faces for Recognition at Scale. *arXiv* **2015**, arXiv:1512.00596. [1512.00596].
22. Liu, H.; Lu, J.; Feng, J.; Zhou, J. Label-Sensitive Deep Metric Learning for Facial Age Estimation. *IEEE Trans. Inf. Forensics Secur.* **2018**, *13*, 292–305. [CrossRef]
23. Howard, A.G.; Zhu, M.; Chen, B.; Kalenichenko, D.; Wang, W.; Weyand, T.; Andreetto, M.; Adam, H. MobileNets: Efficient Convolutional Neural Networks for Mobile Vision Applications. *arXiv* **2017**, arXiv:1704.04861. [1704.04861].
24. Huang, G.; Liu, Z.; Weinberger, K.Q. Densely Connected Convolutional Networks. *arXiv* **2016**, arXiv:1608.06993. [1608.06993].
25. Zhang, Y.; Liu, L.; Li, C.; Loy, C.C. Quantifying Facial Age by Posterior of Age Comparisons. In Proceedings of the British Machine Vision Conference (BMVC), London, UK, 4–7 September 2017.
26. Yang, T.Y.; Huang, Y.H.; Lin, Y.Y.; Hsiu, P.C.; Chuang, Y.Y. SSR-Net: A Compact Soft Stagewise Regression Network for Age Estimation. In Proceedings of the Twenty-Seventh International Joint Conference on Artificial Intelligence, IJCAI-18, Stockholm, Sweden, 13–19 July 2018, pp. 1078–1084.

Article

A Coarse-to-Fine Approach for 3D Facial Landmarking by Using Deep Feature Fusion

Kai Wang [1], Xi Zhao [2,*], Wanshun Gao [1] and Jianhua Zou [1]

1 School of Electrical and Information Engineering, Xi'an Jiaotong University, Xi'an 710049, China; wk19890107@stu.xjtu.edu.cn (K.W.); g-wanshun@stu.xjtu.edu.cn (W.G.); jhzou@sei.xjtu.edu.cn (J.Z.)
2 School of Management, Xi'an Jiaotong University, Xi'an 710049, China
* Correspondence: zhaoxi1@gmail.com

Received: 11 June 2018; Accepted: 24 July 2018; Published: 1 August 2018

Abstract: Facial landmarking locates the key facial feature points on facial data, which provides not only information on semantic facial structures, but also prior knowledge for other kinds of facial analysis. However, most of the existing works still focus on the 2D facial image which may suffer from lighting condition variations. In order to address this limitation, this paper presents a coarse-to-fine approach to accurately and automatically locate the facial landmarks by using deep feature fusion on 3D facial geometry data. Specifically, the 3D data is converted to 2D attribute maps firstly. Then, the global estimation network is trained to predict facial landmarks roughly by feeding the fused CNN (Convolutional Neural Network) features extracted from facial attribute maps. After that, input the local fused CNN features extracted from the local patch around each landmark estimated previously, and other local models are trained separately to refine the locations. Tested on the Bosphorus and BU-3DFE datasets, the experimental results demonstrated effectiveness and accuracy of the proposed method for locating facial landmarks. Compared with existed methods, our results have achieved state-of-the-art performance.

Keywords: facial landmarking; 3D geometry data; 2D attribute maps; fused CNN feature; coarse-to-fine

1. Introduction

Accurate and automatic facial landmark detection or face alignment is critical in face verification, face recognition, facial animation, facial expression recognition and other research. Therefore, it attracts increasing research interests worldwide.

Recently, most studies on face alignment are still primarily conducted on texture images [1–10]. As known, 2D face images are rather sensitive to some condition changes such as arbitrary pose and illumination variations. To address the pose limitation, some researchers proposed that using the reconstructed 3D shape can assist facial landmarking performance under arbitrary poses [11,12]. However, the reconstructed 3D face shape based on corresponding 2D face texture is still sensitive to illumination changes. Motivated by this challenge, the emergence of 3D facial data has provided an alternative to enhance the accuracy and efficiency of facial landmarks' estimation.

With the progress of 3D technology, locating facial landmarks on the 3D facial data has been widely studied [13–21]. Unlike 2D images, both facial geometry information and texture information is contained in each piece of 3D facial data. During the past decade, more studies about facial landmarks' estimation on 3D facial data have been presented. Most of the approaches [20–22] applied both texture data and geometry data to detect landmarks jointly, which can enhance the performance effectively. In fact, not all 3D scanners provide texture and the texture information is not invariant to viewpoint and lighting conditions, so it is necessary to locate landmarks accurately only from 3D geometry data. However, most studies only take range data into account and don't make the best of features extracted from 3D geometry data. In contrast, Li [23] employs feature fusion to recognize facial expression and make great progress. Motivated by this, our proposed method would take five facial attribute maps extracted from 3D geometry data, instead of only applying the range data.

In this paper, we proposed a general framework based on coarse-to-fine for face landmarking only taking 3D facial geometry data. As Figure 1 illustrates, we firstly proposed five feature maps computed from pre-processed 3D geometry data, including a range map, three surface normal maps and a curvature map, which are insensitive to lighting conditions. To locate landmarks accurately, a cascade regression network was designed to update landmarks location iteratively. For this purpose, the global CNN feature extracted by a pre-trained deep neural network from five feature maps was used to estimate landmarks roughly. According to learning the mapping functions from the fused local CNN feature around the landmark estimated previously to corresponding residual distance, local refinement nets are trained independently. By adopting the coarse-to-fine strategy, the performance of landmarking would be improved iteratively.

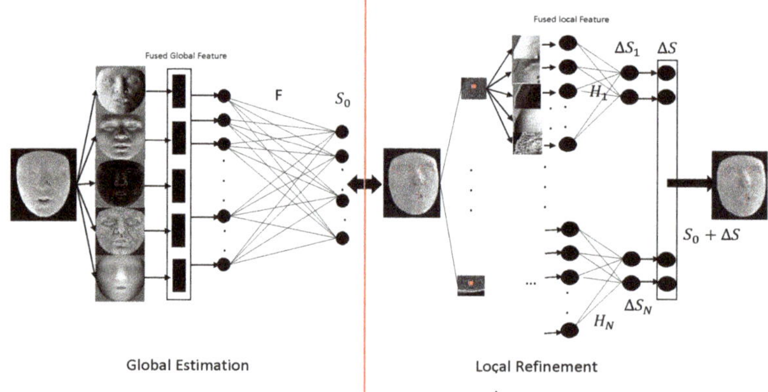

Figure 1. Flowchart of our algorithm for landmarks' detection on 3D facial geometry.

In summary, our learning-based framework is a novel coarse-to-fine approach to estimate landmarks on 3D geometry data by fusing the deep CNN features. The main contributions of this work are the following:

- We propose using the deep CNN feature extracted from five kinds of facial attribute maps to estimate 3D landmarks jointly, instead of using any handcrafted features.
- We propose a global estimation stage and a local refinement stage for 3D landmarks' prediction based on coarse-to-fine strategy and feature fusion.
- Tested in the public 3D face datasets named Bosphrous and BU-3DFE databases, the performances have been state-of-the-art.

The rest of this paper is organized as follows: Section 2 briefly reviews related works about 2D and 3D landmarks' localization. Section 3 describes our proposed method in detail. In this section, the architecture of proposed model, global estimation and local refinement will be introduced. Experimental results are evaluated and compared in Section 4. The weakness of the proposed approach will be discussed in Section 5. Section 6 includes the conclusions and future research derived from this work.

2. Related Work

2.1. Facial Landmarking on 2D Images

Various 3D based methods are the extension of 2D-based. The 2D facial landmarking can generally be divided into two main categories: model-based [1–3] and regression-based [4,6,7] methods. In the former category, it mainly builds face templates to fit the input images, such as Active Appearance Model (AAM) [1], Active Shape Model (ASM) [2], and Constrained Local Model (CLM) [3]. However, model-based methods do not perform not very well in the wild, mainly because the linear model can't handle the complex nonlinear model well. Thus, the regression-based method was proposed to estimate landmark locations explicitly by regression models. It also has been the most widely employed and has made great progress. Supervised Descent Regression (SDR) [6], Cascade Fern Regression (CFR) [7], and Random Forest Regression (RFR) [4] have been established to deal with face alignment on 2D face images. However, most regression-based methods [5,8–10] refine an initial landmark location iteratively, and the performance under some challenging conditions such as illumination changes are not very satisfactory.

Recently, research on deep learning has become a popular field of study with the development of computer hardware and the theory of neural networks. Face recognition [24,25], face verification [26] and facial expression recognition [27] have achieved better performance than the traditional approaches. Compared with the traditional methods, deep learning-based methods have been emerging as an innovative branch in facial landmarking studies recently. Cascade CNN [28], coarse-to-fine Auto-encoder Networks (CFAN) [29] and deep multi-task [30] learning methods are proposed to locate landmarks accurately. Stacked hourglass networks [31] are proposed to estimate landmarks end-to-end. In essence, deep-learning based methods are still regression-based methods which adopt deeper neural networks to estimate the nonlinear correlation between facial image and estimated landmarks. However, it is a great challenge to acquire a huge amount of face data and corresponding labels. Some methods are built on three-dimensional assistance. In Zhu [11], Jourabloo [12] and Kumar [32], they all adopt a 3D solution in a novel alignment framework, which shows that the character of 3D data can help to conquer the limitation of arbitrary pose and other challenges. In Bulat [33], they created a large dataset and estimated 2D and 3D landmarks by adopting hourglass networks. However, all of these methods obtain corresponding 3D shape by adopting 3DMM or 2D texture images that is also sensitive to the changeable lighting conditions.

2.2. Facial Landmarking on 3D Facial Data

Many studies on face landmarking based on 3D geometry and texture data jointly have been proposed recently.

In most of the existing works on 3D facial landmarking, 3D facial landmarks are estimated by computing the 3D shape-related feature, including shape index [14,15,34], effective energy [16], Gabor filter [17,18], local gradient [35] and curvature feature [36]. However, the accuracy on these prominent landmarks decreases drastically, including nose tip and the corner of eyes.

Among these methods on 3D facial landmarking, many approaches utilize registered range data and texture images jointly to estimate landmarks straightforwardly, which can take full advantage of the information from range and texture data. In Boehnen and Russ [37], the eye and mouth maps are computed by adopting both range and texture information. In Wang et al. [38], a point signature representation and the Gabor jets from 2D texture images are used to represent the 3D face mesh. Salah and Jahanbin et al. [22,39] proposed the Gabor wavelet coefficient so that the local appearance in 2D texture image and local patch in the range data around each landmark can be modeled well. As the same thought, in Lu and Jain [40], the local shape index feature and cornerness texture feature around seven landmarks were computed and fused to detect landmarks jointly.

Unlike the above approaches which estimate each landmark independently, the combination of candidate landmarks is quite essential to improve the performance. To make use of the structure between each landmark, the heuristic model [21], 3D geometry-based model [37] and elastic bunch graph-based model [22] were proposed. Most of the works constructed the average 3D position of landmarks as the initialization shape and then updated the position iteratively. However, all of these approaches didn't consider the relationship between the 3D position of landmarks and the feature around each landmark, including the range feature and texture feature. In addition, the 3D point distribution model (PDM) was proposed to estimate eyes, nose and mouth corner. Nair and Cavallaro [21] study 3D facial landmarking by building a statistical model to estimate landmarks coarsely, and then heuristics are applied to refine the locations. Perakis et al. [14,15] study landmarking on 3D facial data under much more challenging conditions, such as the missing data caused by self occlusion. Zhao et al. [20] proposed another method based on statistical models, who presented a model which take the both the relationship between each landmark and the local properties around each landmark into account. However, the main problem of this approach is that the solution is not global, which was caused by the inappropriate initialization.

3. Methodology

3.1. Overview

Given a 3D facial geometry data G, 3D facial landmarks' detection is the task to locate N pre-defined fiducial points, including eye corners, nose tip, mouth corners and so on. We denote the homogeneous coordinate of 3D facial landmarks as S:

$$S = \begin{pmatrix} x_1 & x_2 & \dots & x_N \\ y_1 & y_2 & \dots & y_N \\ z_1 & z_2 & \dots & z_N \end{pmatrix}, \tag{1}$$

where N is the pre-defined number of landmarks. The function is also equal to the following function:

$$S = \begin{pmatrix} x(u_1, v_1), & x(u_2, v_2), & \dots & x(u_N, v_N) \\ y(u_1, v_1), & y(u_2, v_2), & \dots & y(u_N, v_N) \\ z(u_1, v_1), & z(u_2, v_2), & \dots & z(u_N, v_N) \end{pmatrix}, \tag{2}$$

where x,y and z represent the x,y,z coordinate map for each pair (u, v). Given 3D facial data, our goal is to simultaneously estimate the (u, v) accurately.

For this purpose, we propose transforming the 3D face landmarks' estimation to detect the landmarks on five types of 2D facial attribute maps, including shape index map, normal maps and original range map that calculated on 3D geometry data. Then, a novel framework as Figure 1 was presented to achieve our goal accurately and efficiently. Based on the coarse-to-fine strategy, the framework comprises two main parts: one is for global estimation and the other is for local refinement. Specifically, the global estimation phase is intended to locate the landmarks roughly by feeding into the fused global feature that extracted from these attribute maps. Then, the local refinement stage is to learn the nonlinear mapping function from the fused local feature that extracted from a local patch around estimated global landmarks to residual distance.

In the global estimation phase, the goal is to locate landmarks roughly, but it is still more robust and accurate than the mean shape. To train this model, instead of applying the handcrafted feature, we use the pre-trained deep network to extract features from each facial attribute map as a global feature and then concatenate them as the fused feature. Feeding into the fused feature, the target of the regression model is to estimate global landmarks directly. According to the trained model, the global landmarks would be obtained roughly but robustly, which can lay the foundation for the local refinement.

After global optimization by inputting the fused global feature, we can get the initialization shape. The initialization shape is more robust and accurate than the mean shape; however, it is still not satisfied. To refine the global estimation, the refinement stage is designed to refine the results. We extract the local CNN feature from the cropped local patches around the global landmarks and then learn the mapping function from the fused local feature to the residual distance between previous landmarks and ground truth.

3.2. *Facial Attribute Maps*

To comprehensively describe the geometric information of 3D data, five types of facial attribute maps were constructed, including three surface normal maps Nx, Ny, Nz, curvature feature SI, and range data R. Among these maps, surface curvature and normal maps are the most significant feature in 3D object detection, recognition and other 3D tasks. Figure 2 shows the five types of facial attribute maps computed from original 3D facial geometry data.

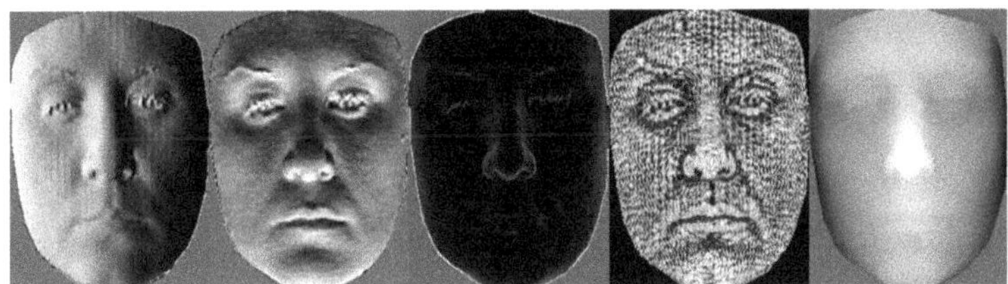

Figure 2. These five facial attribute maps, denoted as three surface normal map N_x, N_y, N_z, curvature feature map SI and range map R.

3.2.1. Surface Curvature Feature

The surface curvature features have been adopted for 3D face landmarks' estimation in many types of research. Actually, surface curvature is the most significant feature in 3D object detection, recognition and other 3D tasks. Thus, this paper chooses the shape index feature map as the first facial attribute.

The Shape index is a continuous mapping of principal curvature values (k_{max}, k_{min}) of a 3D object point p. Once we have two principal curvature (k_{max}, k_{min}), the shape index values, which describe different shapes classed as single numbers ranging from 0 to 1, are calculated as:

$$SI(p) = \frac{1}{2} - \frac{1}{\pi} arctan(\frac{k_{max} + k_{min}}{k_{max} - k_{min}}). \tag{3}$$

3.2.2. Surface Normal Maps

Considering a normalized 3D facial geometry data G, denoted as a $m \times n \times 3$ matrix:

$$G = [P_{uv}(x, y, z)]_{m \times n} = [p_{uvx}, p_{uvy}, p_{uvz}]_{1 \leq u \leq m, 1 \leq v \leq n}, \tag{4}$$

where $[P_{uv}(x, y, z)]$ denotes the corresponding 3D point coordinate of facial geometry data. The corresponding surface normal maps are represented as:

$$\begin{aligned} N(I_g) &= N[P_{uv}(x, y, z)]_{m \times n} \\ &= [N(p_{uvx}), N(p_{uvy}), N(p_{uvz})]_{1 \leq u \leq m, 1 \leq v \leq n}. \end{aligned} \tag{5}$$

In this paper, a local plane fitting method is applied to compute $N(I_g)$, which consists of a three $M \times n$ matrix. In other words, for each point in 3D facial geometry data, the surface normal vector can be computed by the following function:

$$S_{uv} : N_{uvx} q_{uvx} + N_{uvy} q_{uvy} + N_{uvz} q_{uvz} = d, \tag{6}$$

where $(q_{uvx}, q_{uvy}, q_{uvz})$ represents any point within the local neighbourhood of point p_{uv} and $\left\|(N_{uvx}, N_{uvy}, N_{uvz})^T\right\|_2 = 1$. In this paper, a neighbourhood of 5×5 window is adopted and three normal maps would be obtained, denoted as N_x, N_y, N_Z.

3.3. Global Estimation

As the proposed method illustrates, these five types of attribute maps as Figure 2 would be fed into the neural network to estimate landmarks roughly. Considered the calculated feature maps, denoted as shape index SI, N_x, N_y, N_z and original range map R, $S_g(x) \in R^{2N \times 1}$ represents the ground truth of N landmarks. The goal of our global model is to learn the mapping function F from our fused feature map to the ground truth coordinate:

$$S_g(x) \leftarrow F(SI, N_x, N_y, N_z, R). \tag{7}$$

Limited to the amount of training data, training a global CNN model directly is always over-fitting. To overcome this limitation, fine-tuning based on a pre-trained deep model was employed to learn F. To achieve this goal, the parameters of pre-trained model were fixed except training the last layer. Then, the SI, N_x, N_y, N_z, R are fed into the pre-trained model (e.g., VGG (Visual Geometry Group)-net in this paper) separately. Generally, the pre-trained deep CNN model can be regarded as a special feature extractor, which can be regarded as $v = DNN(Map)$, where DNN represents the fixed part of the pre-trained model, Map denotes the resized facial attribute map, and v is the extracted feature vector of each attribute map. Consider adopting shape index maps and convolution neural networks to detect a coarse S_0 as the result of the first step. In particular, the deep models are all comprised of three main parts including convolutional layers, pooling layers and fully connected layers.

- Convolutional Layer and ReLU Non-linearity.

Through a set of designed and learnable filters, the convolutional layer transforms the input images or activation maps to another. Specifically, given a set of activation maps from the previous layer $y^{l-1} \in \mathbb{R}^{W_{l-1} \times H_{l-1} \times D_{l-1}}$, and K_l convolutional filters, each with size $W_f \times H_f \times D_{l-1}$, a list of activation maps $y^l \in \mathbb{R}^{W_l \times H_l \times D_l}$ at the layer L will be computed and output. Let this stride be S; then, the $W_l = (W_{l-1} - W_f + 2P)/S + 1$ and $H_l = (H_{l-1} - H_f + 2P)/S + 1$. Then, we add an activation function φ to adjust the result to a nonlinear function. In this paper, rectified linear units (ReLU), denoted as $\varphi(x) = max(0, x)$, is used. Thus, the result of l layer is denoted as:

$$y^l = \varphi(W_l * y^{l-1} + b_l), \tag{8}$$

where b_l denotes the bias term, and $*$ denotes the convolution operator.

- Fully Connected layers.

This layer is used to reshape these feature maps into a vector feature. The hidden layers are fully connected, which means that each unit in a previous layer is connected with each unit in the next layer. Suppose the global network has L convolutional layers in total and so the feature maps in the last convolutional layers are represented as $y^l \in \mathbb{R}^{W_L \times H_L \times D_L}$. Let the $(L+1)$-th layer be the fully connected layer, and the output of layer L be the input of layer $L+1$, with size $y^{L+1} \in \mathbb{R}^K$, where $K = W_L \times H_L \times D_L$. Thus, this layer is equal to:

$$y^{L+1} = reshape(y^L). \tag{9}$$

Then, the next fully connected layer will be:

$$y^{L+2} = \varphi(w^{L+1} \times y^{L+1} + b^{L+1}), \tag{10}$$

where W^{L+1} is the weight value in the $L+1$-th layer and b^{L+1} is the bias term value. φ denotes the tanh activation function. C. Objective function. After feature extraction for each facial attribute map is done separately, the feature vectors are concatenated as $V = \left[v_{SI}, v_{N_x}, v_{N_y}, v_{N_z}, v_R\right]$ to train the global model F. Specifically, by training a designed neural networks, our target has been formulated as solving the objective function:

$$argmin \left\|S_g - F(V)\right\|_2^2, \tag{11}$$

where F is the nonlinear regression function from V to the landmarks S_g, denoted as $F = \sigma\left(W^T V + b\right)$, where σ represents the nonlinear activation function such as sigmoid, tanh and Relu. In this paper, sigmoid function is employed by the final output layer to learn the parameters $[W, b]$. However, the range of final output is $[0, 1]$, while the range of regression is inconsistent. Therefore, S_g would be normalized to range $[0, 1]$, so that the objective function can be formulated as minimizing the function:

$$argmin \left\|S_g - F(V)\right\|_2^2 + \lambda \left\|W\right\|_F^2, \tag{12}$$

where $\|W\|_F^2$ denotes the regularization term, added to prevent the over-fitting. λ is the set to 0.00005.

After the optimization with Equation (12), the learned parameters $[W, b]$ are obtained and S_0 would be calculated via $S_0 = F(V)$.

3.4. Local Refinement

The global estimation phase describes the mapping function from the fused facial attribute maps to the target landmarks' location. Unlike other methods, the estimated shape is global and more accurate than the mean shape. However, it is still rough and there is room for improvement. To achieve more accurate locations, a coarse-to-fine based approach is proposed to improve the performance. Similar to many cascade regression methods for 2D face alignment, a local model as Figure 3 is employed to estimate the residual distance ΔS, representing the distance between global estimated shape S_0 and ground truth S_g.

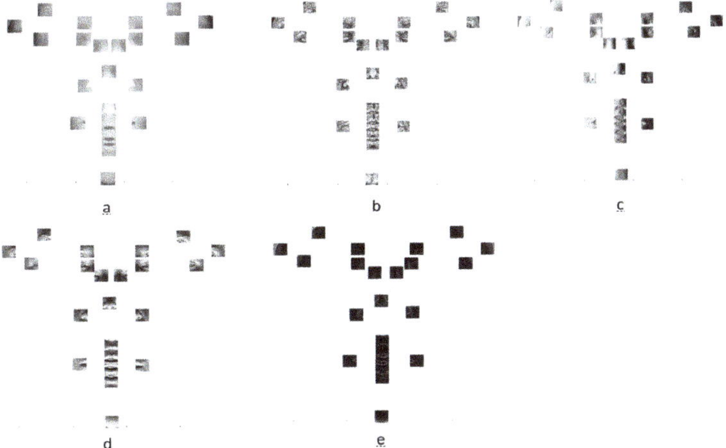

Figure 3. Five different local attribute maps for 22 landmarks. (**a**): depth feature map; (**b**): curvature feature; (**c**): surface normal feature along the x-axis; (**d**): surface normal feature along the y-axis; (**e**): surface normal feature along the z-axis.

Similar to the global estimation, we employed the pre-trained CNN model to extract local features from the local patches around the estimated shape S_0. Each local patch around S_0 is cut out within 30 mm, and then transformed to attribute maps. After the calculation of local attribute maps, they would be resized to 224×224 and are fed into the pre-trained deep neural network to extract local CNN features. Actually, we once considered concatenating the fused local feature of all landmarks to estimate the ΔS jointly. However, limited to the huge number of trained parameters (e.g., $4096 \times 5 \times 22 \times 44 = 19{,}824{,}640$), we propose refining each local patch around a landmark independently. For this purpose, deep feature fusion is also applied for training local model, denoted as $\phi_i = [\phi^i_{SI}, \phi^i_{Nx}, \phi^i_{Ny}, \phi^i_{Nz}, \phi^i_{R}]_{i=1,2,\ldots N}$, where i represents the i-th landmark and N is the number of located landmarks.

Getting the local feature vectors, the local refinement model is to learn a nonlinearity function H_i from fused local feature ϕ_i to the ΔS_i for each landmark, denoted as $\Delta S_i = S_g(i) - S_0(i)$. The objective function of each model can be formulated as follows:

$$argmin \, \|\Delta S_i - H_i\,(\phi_i)\|_2^2 + \beta\,\|W_k\|_F^2, \tag{13}$$

where H_i is a regression function the same as F, represented as $H_i = \sigma\,(W_i\phi_i + b_i)$. Different from the global estimation, the activation function σ is the tanh function, so that all the outputs are in range $[-1, 1]$. After optimization, we can compute ΔS_i according to $\Delta S_i = H_i\,(\phi_i)$, and then we obtain $\Delta S = [\Delta S_1, \Delta S_2, \ldots, \Delta S_N]$. Therefore, normalized results S_{final} can be computed as the following:

$$S_{final} = \Delta S + S_0. \tag{14}$$

4. Experiments

We firstly introduce the datasets used in this paper and then will describe data pre-processing, data augmentation and the parameters' setting briefly in this section. Finally, we will evaluate the performance in these datasets and compare their performances with other methods.

4.1. Datasets

To evaluate the proposed approach, we employ two public 3D facial data, namely the Bosphorus database [41] and the BU-3DFE (Binghamton University 3D Facial Expression) database [42].

The Bosphorus database contains 4666 pairs facial scans from 105 subjects. It also contains 3D facial geometry data under various occlusions (e.g., glass, hands and hair) and several facial expressions. In our experiments, all of the nearly frontal facial data are selected regardless of the occlusion and expressions, resulting in 3632 3D facial geometry data in total. However, the number of landmarks in these data is inconsistent, so we manually selected and labelled 22 landmarks in the Bosphorus dataset for training the models.

The BU-3DFE database includes data from 100 subjects which contain 56 female and 44 male. Each subject contains not only a neutral expression but also the six universal expressions. In our experiments, we have selected all near frontal facial data from all the subjects, regardless of the expression variance, getting 2500 facial scans totally. In this dataset, among the labelled 83 landmarks, we manually selected 68 landmarks and abandoned the other 15 landmarks located on the facial edge. Actually, some common landmarks are labelled in the two datasets, such as eye corners and mouth corners.

4.2. Data Pre-Processing

To learn the global and local attribute maps, the size of global and local patches needed to be resized to the same size, meaning that the number of 3D clouds for each piece of 3D facial geometry data is uniform. However, it is hard to be normalized because of the different face scales. Therefore, uniform grids are applied to remesh the global facial scans or local regions around landmarks. To get local regions, we select all of the points around the landmark with a specific size of 30 mm × 30 mm, and then remesh a uniform grid with the same number of points by using the interpolation. At the same time, the z-values on this grid would be processed by using this normalization. Based on the uniform grids, the facial attribute maps and local patches would be constructed easily and efficiently.

4.3. Data Augmentation

In fact, the number of training data in these datasets is not enough to avoid over-fitting. To overcome over-fitting and improve the performance, increasing the number of training data by utilizing data augmentation is necessary and useful. For this purpose, randomly rotation and symmetry transformation were chosen to augment the variety of facial data. Firstly, we randomly rotate facial data in the horizontal direction and ensure that the face is nearly frontal. Secondly, we also transform the symmetry data for each piece of training data. After data augmentation, more artificially generated facial data would be obtained, so that the over-fitting can be addressed effectively. Of course, the corresponding ground truth would be changed by the same rules.

4.4. Experimental Setting

In our paper, the pre-trained deep CNN model, namely VGG16 [43], is selected for extracting deep CNN features. In the pre-trained networks, all layers and parameters are kept unchanged in the network except the final fully connected layer. As known, the size of the input map is 224 × 224 and the dimension of features is 4096. Since we have five types of facial, the dimension of fused feature is 4096 × 5, while the number of output units is $2 \times N$. The weight matrix W with size $(4096 \times 5) \times (2 \times N)$ would be randomly initialized, and corresponding bias vector b would be initialized by a $2 \times N$-dimensional zero vector. Each local refinement network is almost similar to the global estimation network, and the number of output units is 2. The weight matrix W_i with size $(4096 \times 5) \times 2$ would be also randomly initialized, and the corresponding bias vector b_i would be initialized by a two-dimensional zero vector.

4.5. Convergence and Model Selection

To train these models appropriately, we trained the global estimation model and local refinement models for 2000 iterations, so that these models can converge. Actually, these models have been in convergence when the models were trained about for 1600 iterations. However, to avoid over-fitting in these testing data, the models which trained for about 1400 iterations would be chosen, which may be closed to convergence and more suitable in the testing dataset. The experiments also show that these models perform much better in the testing data.

4.6. Evaluation

To evaluate our proposed approach, three comparison experiments are designed in this section. First, it is necessary to confirm the efficiency of coarse-to-fine strategy. Second, the performance by using mean shape as initialization shape is evaluated. Furthermore, the third is to show the performances under different feature combination. In all experiments, distance error calculated as Euclidean distance between estimated landmarks location and corresponding ground truth were used to evaluate the performance. To evaluate and compare these methods, these three main experiments are carried out on the Bosphorus dataset. Among these 3632 data, 2800 data are randomly selected as training data, and the other 832 are regarded as testing data. The number of training data is increased to $2800 \times 6 = 16{,}800$ after augmentation. In this section, all models are trained and tested by using the same training and testing data.

To confirm the effective of global estimation, we compare our method with the method by taking mean shape as initialization shape. Different from taking the global estimation as initialization, mean shape is computed as the initialization shape for local refinement. Instead of global estimation, the local patches around mean shape are taken to extract local features. Then, we will update the locations the same as the local refinement phase in our method. Figure 4a shows the average distance error after global estimation and mean shape calculation, and Figure 4b illustrates the average distance error via two different initialization ways after local refinement. As can be seen, the results of our proposed method outperforms after the local refinement.

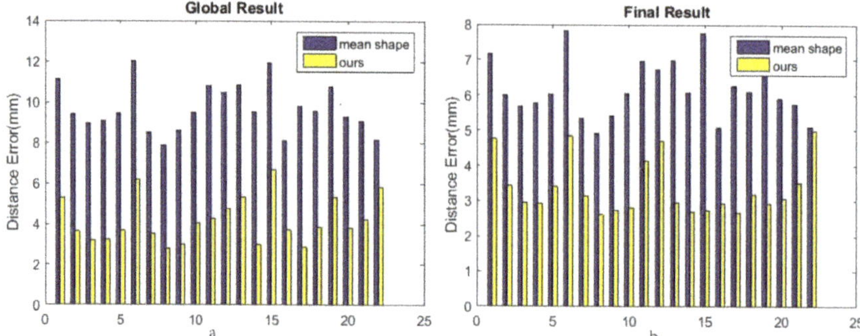

Figure 4. The comparison results between mean shape and our proposed method. (**a**) denotes the results after global estimation and the (**b**) represents the results after refinement.

Furthermore, to verify the coarse-to-fine strategy, we compare the results after global estimation and local refinement. In Figure 5, the blue bars show the average distance error of 22 landmarks in the testing dataset after global estimation, while the other bars show the results after refinement. It can be easily observed that the results are enhanced effectively from coarse to fine. Note that the mean error has achieved 4.11 mm after global estimation, while 98.23% landmarks are located automatically with 20 mm and 93.31% landmarks are with 10 mm. After local refinement, the 100% landmarks are located automatically with 20 mm precision and 96.43% are with 10 mm. Furthermore, the average error of all landmarks in the testing data can also be improved to 3.37 mm, which has achieved the state-of-the-art.

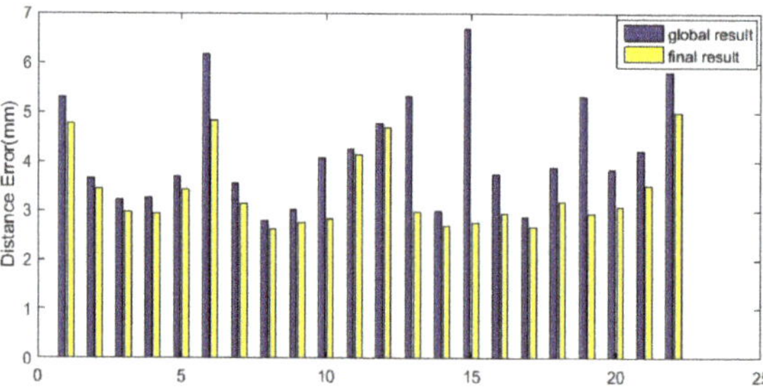

Figure 5. The comparison results after global estimation and local refinement.

To show the performance under different feature combinations, the experiment is carried on the same training and testing data, and independent models are trained under different feature combinations. For this purpose, we selected maps from five facial attribute maps randomly and $30 = (2^5 - 2)$ kinds of feature combinations are generated to train and test models separately. In the case of each condition, the number of inputs would be modified to adjust the different network architecture, and other parameters in the networks are invariable. Figure 6 shows the global estimation results under different feature combinations. In this figure, the blue bars represent the mean error when different feature sets are fed into the network, while the red bar denotes our result. It can be observed that our global estimation result is the best, especially when we fuse all of these five facial attribute maps.

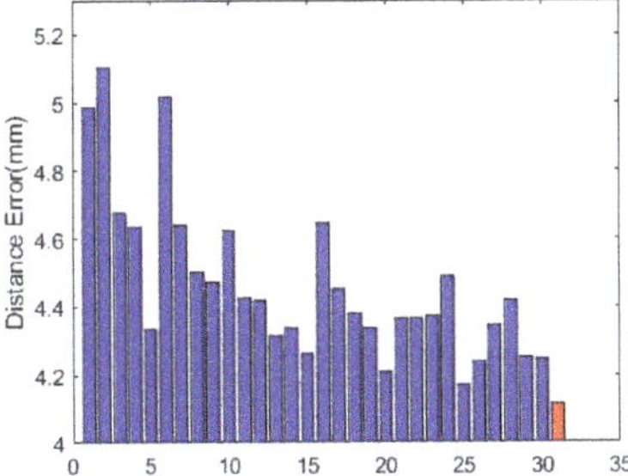

Figure 6. The global estimation results under different feature fusion.

4.7. Comparison with Other Methods

4.7.1. Comparison with Handcrafted Features

To compare the performance of deep fusion feature with the results obtained by applying handcrafted features, their handcrafted features were tested. Instead of the deep fusion feature, three classical features including HOG (Histogram of Oriented Gradient), SIFT (Scale Invariant Feature Transform) and LBP (Local Binary Pattern), which have been proved to be efficient for image analysis, were employed to locate landmarks iteratively. For this purpose, these features around mean shape are firstly extracted and then respectively fused and fed into the designed networks to estimate landmarks coarse-to-fine with default parameters. Table 1 shows the average location error across all of the 22 landmarks on the Bosphporus database. We can easily draw the conclusion that the deep feature fusion marked with the bold fonts based on the pre-trained model is more accurate than the handcrafted features for all of these 22 landmarks. Furthermore, among these handcrafted features, the SIFT feature achieves the best performance, and outperforms HOG and LBP. These results also indicate that the location performance would obviously be affected by different features.

Table 1. Comparison with hand-crafted features on the Boshporus database.

Landmarks	SIFT	LBP	HOG	Deep Features
Outer left eyebrow	6.13 ± 3.97	6.45 ± 4.11	6.38 ± 4.37	**4.76 ± 3.15**
Middle left eyebrow	5.37 ± 2.15	4.95 ± 2.07	5.68 ± 3.62	**3.43 ± 2.38**
Inner left eyebrow	5.14 ± 3.23	5.28 ± 3.45	5.48 ± 2.08	**2.96 ± 2.14**
Inner right eyebrow	5.04 ± 2.78	5.18 ± 2.96	5.34 ± 3.05	**2.93 ± 1.79**
Middle right eyebrow	4.88 ± 2.86	5.03 ± 2.54	5.08 ± 2.86	**3.41 ± 2.06**
Outer right eyebrow	6.02 ± 3.50	5.97 ± 3.45	6.17 ± 3.74	**4.83 ± 4.07**
Outer left eye corner	4.16 ± 2.05	4.83 ± 2.36	4.97 ± 2.60	**3.14 ± 2.17**
Inner left eye corner	4.53 ± 2.53	4.12 ± 2.27	5.02 ± 3.10	**2.62 ± 1.73**
Inner right eye corner	3.71 ± 2.19	4.03 ± 2.30	4.34 ± 2.62	**2.74 ± 1.24**
Outer right eye corner	4.09 ± 2.51	3.89 ± 2.84	4.13 ± 2.74	**2.82 ± 1.85**
Nose saddle left	7.85 ± 4.03	7.71 ± 3.96	7.91 ± 4.07	**4.13 ± 2.75**
Nose saddle right	8.23 ± 4.29	8.35 ± 4.02	8.41 ± 4.72	**4.69 ± 3.18**
Left nose peak	3.54 ± 2.06	3.67 ± 2.17	3.97 ± 2.37	**2.96 ± 2.24**
Nose tip	3.84 ± 2.43	3.91 ± 2.59	4.01 ± 2.77	**2.69 ± 1.95**
Right nose peak	3.53 ± 2.34	3.81 ± 2.61	3.48 ± 2.22	**2.74 ± 2.27**
Left mouth corner	4.39 ± 2.82	4.13 ± 2.58	4.47 ± 3.01	**2.93 ± 3.24**
Upper lip outer middle	4.73 ± 3.12	4.99 ± 3.19	4.45 ± 3.08	**2.66 ± 2.63**
Right mouth corner	6.32 ± 3.83	6.41 ± 3.95	7.04 ± 4.37	**3.18 ± 2.93**
Upper lip inner middle	4.86 ± 2.75	4.64 ± 2.67	4.93 ± 3.15	**2.92 ± 2.65**
Lower lip inner middle	5.15 ± 5.02	5.61 ± 4.96	5.89 ± 5.12	**3.07 ± 3.17**
Lower lip outer middle	6.19 ± 4.19	6.20 ± 3.95	6.07 ± 4.12	**3.51 ± 3.15**
Chin middle	7.69 ± 5.39	7.93 ± 5.62	8.01 ± 5.70	**4.99 ± 4.16**
Mean error	5.25 ± 3.18	5.32 ± 3.21	5.51 ± 3.43	**3.37 ± 2.72**

4.7.2. Comparison with Pre-Trained Models

This section compares the performance of deep fused features based on three different pre-trained models on the ImageNet dataset [43–45]. As aforementioned, different features extracted by using different pre-trained models were fed into the coarse-to-fine networks separately. In this paper, the same as the other handcrafted features, we use these pre-trained models to extract features from these facial attribute maps independently and fuse these features to train the designed model. Limited to numbers of the data, we keep all parameters fixed except the last fully connected layer. We only tested three classical deep models, including AlexNet [44], VGG-net [43] and Google Inception [45]. Table 2 shows the average location errors across all of the 22 landmarks on the Bosphorus database. The best performance is marked by bold fonts. From it, we can conclude that: (1) all of the deep features achieve better performance than the handcrafted features; (2) Deep fusion features all can achieve satisfied performance; and the (3) Google Inception network and AlexNet outperform the VGG-net for a few landmarks. However, comparing with VGG-net, Inception net takes too much time to extract features because of the complex architecture, and AlexNex is unsatisfactory among most of landmarks. Considering the computation accuracy and time complexity, the VGG-net has been chosen as the pre-trained deep model.

Table 2. Comparison with pre-trained deep models on BoshporusDB.

landmarks	AlexNet	Google Inception	VGG-Net
Outer left eyebrow	4.93 ± 2.54	**4.47 ± 2.31**	4.76 ± 3.15
Middle left eyebrow	4.19 ± 3.18	3.62 ± 2.47	**3.43 ± 2.38**
Inner left eyebrow	3.05 ± 2.43	**2.88 ± 2.04**	2.96 ± 2.14
Inner right eyebrow	3.16 ± 2.17	3.04 ± 1.92	**2.93 ± 1.79**
Middle right eyebrow	3.61 ± 2.58	3.55 ± 1.99	**3.41 ± 2.06**
Outer right eyebrow	**4.02 ± 4.16**	4.23 ± 4.35	4.83 ± 4.07
Outer left eye corner	3.16 ± 2.00	3.46 ± 2.10	**3.14 ± 2.17**
Inner left eye corner	2.39 ± 1.60	**2.30 ± 1.40**	2.62 ± 1.73
Inner right eye corner	3.10 ± 2.49	2.87 ± 1.54	**2.74 ± 1.24**
Outer right eye corner	3.01 ± 2.05	**2.77 ± 1.94**	2.82 ± 1.85
Nose saddle left	4.61 ± 3.56	4.88 ± 3.67	**4.13 ± 2.75**
Nose saddle right	5.71 ± 4.13	5.30 ± 3.71	**4.69 ± 3.18**
Left nose peak	3.51±2.99	3.11 ± 2.69	**2.96 ± 2.24**
Nose tip	3.31 ± 2.21	3.01 ± 2.07	**2.69 ± 1.95**
Right nose peak	**2.56 ± 2.04**	2.88 ± 2.50	2.74 ± 2.27
Left mouth corner	4.10 ± 3.74	3.43 ± 3.34	**2.93 ± 3.24**
Upper lip outer middle	3.29 ± 3.01	2.97 ± 2.85	**2.66 ± 2.63**
Right mouth corner	4.19 ± 3.45	3.57 ± 3.22	**3.18 ± 2.93**
Upper lip inner middle	3.61 ± 3.42	**2.87 ± 3.15**	2.92 ± 2.65
Lower lip inner middle	4.15 ± 5.04	3.59 ± 4.13	**3.07 ± 3.17**
Lower lip outer middle	4.19 ± 3.89	3.81 ± 3.77	**3.51 ± 3.15**
Chin middle	5.05 ± 5.04	5.13 ± 5.13	**4.99 ± 4.16**
Mean error	3.77 ± 3.08	3.53 ± 2.83	**3.37 ± 2.72**

4.7.3. Comparison on the Bosphorus Dataset

Furthermore, we compared our proposed approach with other existing methods on the Bosphorus dataset. Figure 7 depicts the mean distance error and standard deviation of 22 detected landmarks. From this figure, the mean distance error of all landmarks in the testing data is 3.37 mm, which has achieved the state-of-the-art, especially in some landmarks such as middle left/right eyebrow and so on. Compared with some other existing methods in these common landmarks, the comparison results are shown in Table 3. The best performance is marked by bold fonts. From it, we can see that our approach outperforms in outer eye corners, chin and mouth corners, which are difficult to locate. Figure 8 illustrates some examples of facial landmarking by the proposed approach on this dataset.

In this figure, 3D facial geometry data are rotated through several directions, so that the performance of landmarking can be observed more clearly.

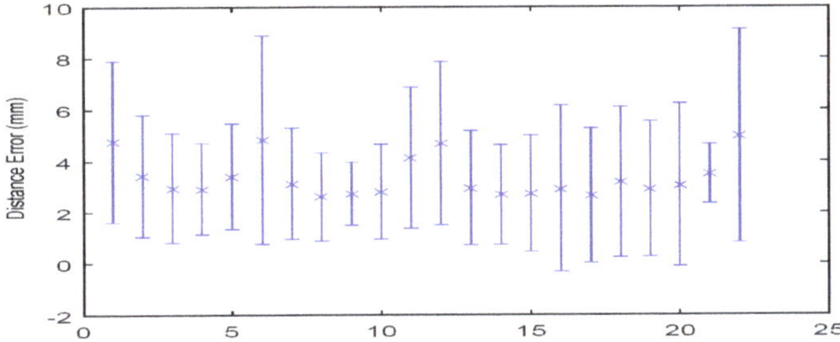

Figure 7. Mean distance error and standard deviation of 22 landmarks on the Bosphorus dataset.

Table 3. Comparison with other methods on BoshporusDB.

	Inner Eye Corners	Outer Eye Corners	Nose Tip	Nose Corners	Mouth Corners	Chin
Manual [46]	2.51	-	2.96	1.75	-	-
Alyuz [46]	3.70	-	3.05	3.10	-	-
Creusot [47]	4.14 ± 2.63	6.27 ± 3.98	4.33 ± 2.62	4.16 ± 2.35	7.95 ± 5.44	15.38 ± 10.49
Sukno [48]	2.85 ± 2.02	5.06 ± 3.67	**2.33 ± 1.78**	**3.02 ± 1.91**	6.08 ± 5.13	7.58 ± 6.72
Camgoz (SIFT) [49]	**2.26 ± 1.79**	4.23 ± 2.94	2.72 ± 2.19	4.57 ± 3.62	3.14 ± 2.71	5.72 ± 4.31
Camgoz (HOG) [49]	2.33 ± 1.92	4.11 ± 3.01	2.69 ± 2.20	4.49 ± 3.62	3.16 ± 2.70	5.87 ± 4.19
Ours	2.66 ± 1.49	**3.64 ± 2.01**	2.69 ± 1.95	4.40 ± 2.61	**3.06 ± 3.09**	**4.99 ± 4.16**

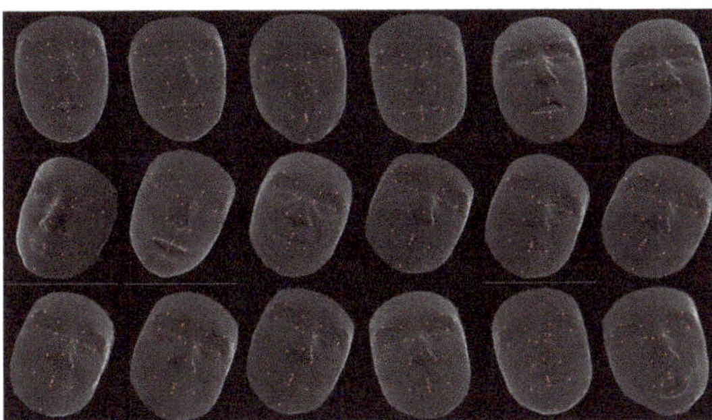

Figure 8. Samples of facial landmarking on 3D facial geometry data on the Bosphorus Dataset. To observe the performance more clearly, we rotate the facial data and estimated landmarks through several directions.

4.7.4. Comparison on the BU-3DFE Dataset

The second experiment is carried out on the BU-3DFE dataset. Among the 2500 facial geometry data, 2000 facial scans from the 100 subjects were selected as the training data. The other 500 facial geometry data were used as testing data. After data argumentation, 12,000 facial scans can be obtained that contain neural expressions and six universal facial expressions. Figure 9 illustrates average distance error and standard deviation of 68 landmarks in the testing dataset of the 68 landmarks. Meanwhile, 98.88% of the landmarks are located with a 20 mm precision, and 93.20% are with the 10 mm precision. The mean distance error of all 68 landmarks has been improved to 4.03 mm. Compared with some other methods in the common landmarks on BU-3DFE dataset, Table 4 depicts the comparison results of 14 common landmarks. The best performance is marked by bold fonts. We can see that the average error of these points has been achieved 3.96 mm and the results in several points outperform, including the outer corner of the left eye, center of the upper lip, and center of the lower lip.

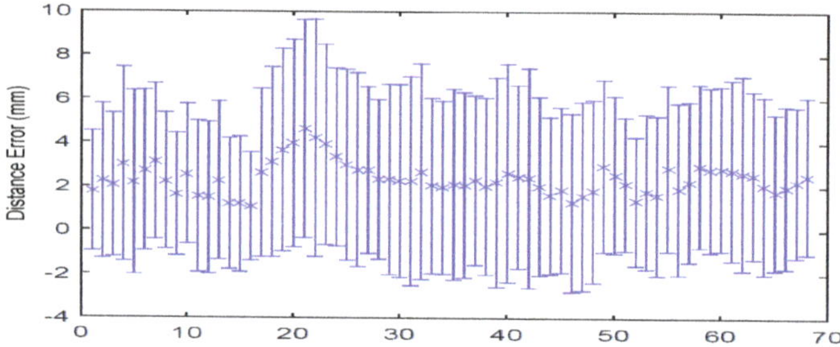

Figure 9. Mean distance error and standard deviation of 68 landmarks on the BU3DFE dataset.

Table 4. Comparison results with existing methods on BU3DFE.

Landmark	Fanelli [50]	Zhao [20]	Nair [21]	Sun [51]	Our Method
Inner corner of left eye	**2.60 ± 1.80**	2.93 ± 1.40	11.89	3.35 ± 5.67	2.79 ± 1.63
Outer corner of left eye	3.60 ± 2.40	4.11 ± 1.89	19.38	3.89 ± 6.38	**3.58 ± 2.27**
Inner corner of right eye	**2.80 ± 2.00**	2.90 ± 1.36	12.11	3.27 ± 5.51	3.11 ± 2.24
Outer corner of right eye	4.00 ± 2.80	4.07 ± 2.00	20.46	**3.73 ± 6.14**	4.20 ± 2.18
Left corner of nose	3.90 ± 2.00	**3.32 ± 1.94**	-	3.60 ± 4.01	3.77 ± 1.87
Right corner of nose	4.10 ± 2.20	3.62 ± 1.91	-	**3.43 ± 3.74**	4.98 ± 2.63
Left corner of mouth	4.70 ± 3.50	7.15 ± 4.64	-	3.95 ± 4.17	**3.88 ± 2.86**
center of upper lip	3.50 ± 2.50	4.19 ± 2.34	-	3.09 ± 3.06	**2.94 ± 1.35**
Right corner of mouth	4.90 ± 3.60	7.52 ± 4.57	-	**3.76 ± 4.05**	3.94 ± 2.96
Center of lower lip	5.20 ± 5.20	8.82 ± 7.12	-	4.36 ± 6.03	**3.73 ± 2.97**
Outer corner of left brow	5.80 ± 3.80	6.26 ± 3.72	-	5.29 ± 6.93	**4.92 ± 2.69**
Inner corner of left brow	**3.80 ± 2.70**	4.87 ± 2.99	-	4.62 ± 5.92	3.81 ± 2.75
Inner corner of right brow	4.00 ± 3.00	4.88 ± 2.97	-	4.59 ± 5.76	**3.85 ± 2.63**
Outer corner of right brow	6.20 ± 4.30	6.07 ± 3.35	-	**5.29 ± 7.04**	5.98 ± 4.63
Mean results	4.22 ± 2.99	5.05 ± 3.01	-	4.02 ± 5.32	**3.96 ± 2.55**

5. Discussion

With the development of deep learning, more and more data is needed to train a robust and accurate model. Unlike 2D images that can be easily obtained from the web, the 3D geometry data can't be constructed easily without professional equipment. Nowadays, the existing 3D geometry databases are all collected from labs and under the controlled conditions. Furthermore, the number of data is far from enough to train an appropriate deep model, so we need to fine-tune the pre-trained model. In this paper, using the pre-trained deep model to extract features from the different attribute maps is essential in the proposed approach. In most of the cases, fine-tuning these deep models means that most of the parameters in the pre-trained models remain unchanged and only a few are updated for specific tasks. For this purpose, we can update the parameters in the last layer or other layers based on the amount of training data. Thus, in our paper, limited to the number of 3D geometry data, we only updated the last layer and didn't test the other choices at all.

In addition, feature fusion is the key step in the proposed approach. Applying the fused feature extracted from deep model can take more useful information into account for locating landmarks. For 3D data, more useful information can be obtained including surface normal, curvature and other attribute maps. In this paper, we only select these five types of attribute maps to train the model. In fact, for each attribute map, the features can be extracted based on different pre-trained models. It is another way to improve the location performance, but it is too complex to be applied in the other testing data satisfied. On the other hand, a classical pre-trained model named ResNet was not considered because of the computational complexity and our computer performance. Although the model would achieve the best performance for our task perhaps, it still cost more than 3 min to extract the features without updating any parameters. For this reason, ResNet was not selected in our approach.

As other research about deep learning, the main weakness is also the computation complexity. Compared with other effective approaches, the computation complexity of our proposed method is higher than the others. In addition, this paper is the first time to utilize the deep-learning based approach to estimate 3D landmarks, while the other effective methods are all based on traditional ways such as hand-crafted features. Actually, to improve the accuracy, higher computation complexity is needed. Benefiting from more and more powerful computing power, the execution time is still satisfied. Of course, a lot of works will be done to reduce the computation complexity and to ensure the accuracy improvement synchronously in future works.

Although our algorithm has achieved state-of-the-art performance, there are a few other works to study. Firstly, we didn't take the profile face into account because there are only a few 3D profile data and fewer landmarks to train a unified location architecture. In addition, data missing caused by posing is the most challenging issue and the main weakness of our algorithm.

6. Conclusions

In this paper, we propose a novel approach to estimate landmarks on 3D geometry data. By transforming the 3D data to 2D attribute maps, the goal of our approach is to predict the landmarks based on the attribute maps. Different from using the handcrafted feature, we feed the global and the local attribute maps into the deep CNN model to extract global and local feature. Based on coarse-to-fine strategy, a global model is trained to estimate landmarks roughly and local models are trained to refine the landmarks' location. Evaluated on the Bosphorus dataset, the proposed method performs more effectively than handcrafted features and other pre-trained models. Compared with other existing methods, the results on the Bosphorus dataset and BU-3DFE dataset have also demonstrated comparable performance, especially in some common landmarks.

In the future, some other issues of improving the robustness under other challenging conditions such as self-occlusion and data missing will be studied. In addition, using decision fusion of simple classifiers to balance the computation complexity and the accuracy may be another effective method for this problem.

Author Contributions: K.W. designed the algorithm, conceived of, designed and performed the experiments, analyzed the data and wrote this paper. X.Z. provided the most important comments and suggestions, and also revised the paper. W.G. and J.Z. provided some suggestions and comments for the performance improvement of the algorithm.

Funding: This research received no external funding.

Acknowledgments: This work was supported by the Natural Science Foundation of China (Grant No. 91746111, Grant No. 71702143), the Ministry of Education and China Mobile Joint Research Fund Program (No. MCM20160302), the Shaanxi Provincial Development and Reform Commission (No. SFG2016789), and the Xi'an Science and Technology Bureau (No. 2017111SF/RK005-(7)).

Conflicts of Interest: The authors declare no conflict of interest.

References

1. Cootes, T.F.; Edwards, G.J.; Taylor, C.J. Active Appearance Models. *IEEE Trans. Pattern Anal. Mach. Intell.* **2001**, *23*, 681–685. [CrossRef]
2. Cootes, T.F.; Taylor, C.J.; Cooper, D.H.; Graham, J. Active shape models—Their training and application. *Comput. Vis. Image Underst.* **1995**, *61*, 38–59. [CrossRef]
3. Cristinacce, D.; Cootes, T.F. Feature Detection and Tracking with Constrained Local Models. In Proceedings of the British Machine Vision Conference 2006, Edinburgh, UK, 4–7 September 2006; pp. 929–938.

4. Kazemi, V.; Sullivan, J. One Millisecond Face Alignment with an Ensemble of Regression Trees. In Proceedings of the IEEE Conference on Computer Vision and Pattern Recognition, Columbus, OH, USA, 23–28 June 2014; pp. 1867–1874.
5. Ren, S.; Cao, X.; Wei, Y.; Sun, J. Face Alignment at 3000 FPS via Regressing Local Binary Features. In Proceedings of the IEEE Conference on Computer Vision and Pattern Recognition, Columbus, OH, USA, 23–28 June 2014; pp. 1685–1692.
6. Xiong, X.; Torre, F.D.L. Supervised Descent Method and Its Applications to Face Alignment. In Proceedings of the IEEE Conference on Computer Vision and Pattern Recognition, Portland, OR, USA, 23–28 June 2013; pp. 532–539.
7. Dollar, P.; Welinder, P.; Perona, P. Cascaded pose regression. In Proceedings of the Computer Vision and Pattern Recognition, San Francisco, CA, USA, 13–18 June 2010; pp. 1078–1085.
8. Savran, A.; Sankur, B.; Bilge, M.T. Regression-based intensity estimation of facial action units. *Image Vis. Comput.* **2012**, *30*, 774–784. [CrossRef]
9. Feng, Z.H.; Huber, P.; Kittler, J.; Christmas, W.; Wu, X.J. Random Cascaded-Regression Copse for Robust Facial Landmark Detection. *IEEE Signal Process. Lett.* **2014**, *22*, 76–80. [CrossRef]
10. Zhu, S.; Li, C.; Chen, C.L.; Tang, X. Face alignment by coarse-to-fine shape searching. In Proceedings of the Computer Vision and Pattern Recognition, Boston, MA, USA, 7–12 June 2015; pp. 4998–5006.
11. Zhu, X.; Lei, Z.; Liu, X.; Shi, H.; Li, S.Z. Face Alignment Across Large Poses: A 3D Solution. In Proceedings of the IEEE Conference on Computer Vision and Pattern Recognition, Las Vegas, NV, USA, 27–30 June 2016; pp. 146–155.
12. Jourabloo, A.; Liu, X. Pose-Invariant 3D Face Alignment. In Proceedings of the IEEE International Conference on Computer Vision, Santiago, Chile, 11–18 December 2016; pp. 3694–3702.
13. Kakadiaris, I.A.; Passalis, G.; Toderici, G.; Murtuza, M.N.; Lu, Y.; Karampatziakis, N.; Theoharis, T. Three-dimensional face recognition in the presence of facial expressions: An annotated deformable model approach. *IEEE Trans. Pattern Anal. Mach. Intell.* **2007**, *29*, 640. [CrossRef] [PubMed]
14. Perakis, P.; Theoharis, T.; Passalis, G.; Kakadiaris, I.A. Automatic 3D facial region retrieval from multi-pose facial datasets. In Proceedings of the Eurographics Conference on 3D Object Retrieval, Munich, Germany, 29 March 2009; pp. 37–44.
15. Perakis, P.; Passalis, G.; Theoharis, T.; Toderici, G.; Kakadiaris, I.A. Partial matching of interpose 3D facial data for face recognition. In Proceedings of the IEEE International Conference on Biometrics: Theory, Applications, and Systems, Washington, DC, USA, 28–30 September 2009; pp. 1–8.
16. Xu, C.; Tan, T.; Wang, Y.; Quan, L. Combining local features for robust nose location in 3D facial data. *Pattern Recognit. Lett.* **2006**, *27*, 1487–1494. [CrossRef]
17. D'Hose, J.; Colineau, J.; Bichon, C.; Dorizzi, B. Precise Localization of Landmarks on 3D Faces using Gabor Wavelets. In Proceedings of the IEEE International Conference on Biometrics: Theory, Applications, and Systems, Crystal City, VA, USA, 27–29 September 2007; pp. 1–6.
18. Colbry, D.; Stockman, G.; Jain, A. Detection of Anchor Points for 3D Face Veri.cation. In Proceedings of the IEEE Computer Society Conference on Computer Vision and Pattern Recognition, San Diego, CA, USA, 20–25 June 2005.
19. Bevilacqua, V.; Casorio, P.; Mastronardi, G. Extending Hough Transform to a Points' Cloud for 3D-Face Nose-Tip Detection. In Proceedings of the International Conference on Intelligent Computing: Advanced Intelligent Computing Theories and Applications—With Aspects of Artificial Intelligence, Shanghai, China, 15–18 September 2008; pp. 1200–1209.
20. Zhao, X.; Dellandréa, E.; Chen, L.; Kakadiaris, I.A. Accurate landmarking of three-dimensional facial data in the presence of facial expressions and occlusions using a three-dimensional statistical facial feature model. *IEEE Trans. Syst. Man Cybern. Part B Cybern.* **2011**, *41*, 1417–1428. [CrossRef] [PubMed]
21. Nair, P.; Cavallaro, A. 3-D Face Detection, Landmark Localization, and Registration Using a Point Distribution Model. *IEEE Trans. Multimedia* **2009**, *11*, 611–623. [CrossRef]
22. Jahanbin, S.; Choi, H.; Jahanbin, R.; Bovik, A.C. Automated facial feature detection and face recognition using Gabor features on range and portrait images. In Proceedings of the IEEE International Conference on Image Processing, San Diego, CA, USA, 12–15 October 2008; pp. 2768–2771.
23. Huibin, L.I.; Sun, J.; Zongben, X.U.; Chen, L. Multimodal 2D+3D Facial Expression Recognition with Deep Fusion Convolutional Neural Network. *IEEE Trans. Multimedia* **2017**, *19*, 2816–2831.

24. Schroff, F.; Kalenichenko, D.; Philbin, J. FaceNet: A unified embedding for face recognition and clustering. In Proceedings of the IEEE Conference on Computer Vision and Pattern Recognition, Boston, MA, USA, 7–12 June 2015; pp. 815–823.
25. Sun, Y.; Liang, D.; Wang, X.; Tang, X. DeepID3: Face Recognition with Very Deep Neural Networks. *arXiv* **2015**, arXiv:1502.00873.
26. Taigman, Y.; Yang, M.; Ranzato, M.; Wolf, L. DeepFace: Closing the Gap to Human-Level Performance in Face Verification. In Proceedings of the Computer Vision and Pattern Recognition, Columbus, OH, USA, 23–28 June 2014; pp. 1701–1708.
27. Chang, F.J.; Tran, A.T.; Hassner, T.; Masi, I.; Nevatia, R.; Medioni, G. ExpNet: Landmark-Free, Deep, 3D Facial Expressions. In Proceedings of the 2018 13th IEEE International Conference on Automatic Face & Gesture Recognition (FG 2018), Xi'an, China, 15–19 May 2018.
28. Sun, Y.; Wang, X.; Tang, X. Deep Convolutional Network Cascade for Facial Point Detection. In Proceedings of the Computer Vision and Pattern Recognition, Portland, OR, USA, 23–28 June 2013; pp. 3476–3483.
29. Zhang, J.; Shan, S.; Kan, M.; Chen, X. Coarse-to-Fine Auto-Encoder Networks (CFAN) for Real-Time Face Alignment. In *European Conference on Computer Vision;* Springer: Cham, Switzerland, 2014; pp. 1–16.
30. Zhang, Z.; Luo, P.; Chen, C.L.; Tang, X. Facial Landmark Detection by Deep Multi-task Learning. In Proceedings of the European Conference on Computer Vision, Zurich, Switzerland, 6–12 September 2014; pp. 94–108.
31. Yang, J.; Liu, Q.; Zhang, K. Stacked Hourglass Network for Robust Facial Landmark Localisation. In Proceedings of the Computer Vision and Pattern Recognition Workshops, Honolulu, HI, USA, 21–26 July 2017; pp. 2025–2033.
32. Kumar, A.; Chellappa, R. Disentangling 3D Pose in A Dendritic CNN for Unconstrained 2D Face Alignment. In Proceedings of the IEEE Conference on Computer Vision and Pattern Recognition, Salt Lake City, UT, USA, 18–22 June 2018.
33. Bulat, A.; Tzimiropoulos, G. How Far are We from Solving the 2D and 3D Face Alignment Problem? (and a Dataset of 230,000 3D Facial Landmarks). In Proceedings of the International Conference on Computer Vision, Venice, Italy, 22–29 October 2017; pp. 1021–1030.
34. Lu, X.; Jain, A.K.; Colbry, D. Matching 2.5D Face Scans to 3D Models. *IEEE Trans. Pattern Anal. Mach. Intell.* **2006**, *28*, 31–43. [PubMed]
35. Dibeklioglu, H.; Salah, A.A.; Akarun, L. 3D Facial Landmarking under Expression, Pose, and Occlusion Variations. In Proceedings of the IEEE International Conference on Biometrics: Theory, Applications and Systems, Arlington, VA, USA, 29 September–1 October 2008; pp. 1–6.
36. Colombo, A.; Cusano, C.; Schettini, R. 3D face detection using curvature analysis. *Pattern Recognit.* **2006**, *39*, 444–455. [CrossRef]
37. Boehnen, C.; Russ, T. A Fast Multi-Modal Approach to Facial Feature Detection. In Proceedings of the Seventh IEEE Workshops on Application of Computer Vision, Breckenridge, CO, USA, 5–7 January 2005; pp. 135–142.
38. Wang, Y.; Chua, C.S.; Ho, Y.K. Facial feature detection and face recognition from 2D and 3D images. *Pattern Recognit. Lett.* **2002**, *23*, 1191–1202. [CrossRef]
39. Salah, A.A.; Çinar, H.; Akarun, L.; Sankur, B. Robust facial landmarking for registration. *Ann. Télécommun.* **2007**, *62*, 83–108.
40. Lu, X.; Jain, A.K. Automatic Feature Extraction for Multiview 3D Face Recognition. In Proceedings of the International Conference on Automatic Face and Gesture Recognition, Southampton, UK, 10–12 April 2006; pp. 585–590.
41. Savran, A.; Akarun, L. Bosphorus Database for 3D Face Analysis. In *Biometrics and Identity Management*; Springer: Berlin/Heidelberg, Germany, 2008; pp. 47–56.
42. Yin, L.; Wei, X.; Sun, Y.; Wang, J.; Rosato, M.J. A 3D facial expression database for facial behavior research. In Proceedings of the FGR'06 International Conference on Automatic Face and Gesture Recognition, Southampton, UK, 10–12 April 2006; pp. 211–216.
43. Simonyan, K.; Zisserman, A. Very Deep Convolutional Networks for Large-Scale Image Recognition. *arXiv* **2014**, arXiv:1409.1556.

44. Krizhevsky, A.; Sutskever, I.; Hinton, G.E. ImageNet classification with deep convolutional neural networks. In Proceedings of the International Conference on Neural Information Processing Systems, Doha, Qatar, 26–29 November 2012; pp. 1097–1105.
45. Ioffe, S.; Szegedy, C. Batch Normalization: Accelerating Deep Network Training by Reducing Internal Covariate Shift. *arXiv* **2015**, arXiv:1502.03167.
46. Alyüz, N.; Gökberk, B.; Akarun, L. Regional registration for expression resistant 3-D face recognition. *IEEE Trans. Inf. Forensics Secur.* **2010**, *5*, 425–440. [CrossRef]
47. Creusot, C.; Pears, N.; Austin, J. A Machine-Learning Approach to Keypoint Detection and Landmarking on 3D Meshes. *Int. J. Comput. Vis.* **2013**, *102*, 146–179. [CrossRef]
48. Sukno, F.M.; Waddington, J.L.; Whelan, P.F. 3-D Facial Landmark Localization With Asymmetry Patterns and Shape Regression from Incomplete Local Features. *IEEE Trans. Cybern.* **2017**, *45*, 1717–1730. [CrossRef] [PubMed]
49. Camgöz, N.C.; Gökberk, B.; Akarun, L. Facial landmark localization in depth images using Supervised Descent Method. In Proceedings of the Signal Processing and Communications Applications Conference, Malatya, Turkey, 16–19 May 2015; pp. 378–383.
50. Fanelli, G.; Dantone, M.; Gool, L.V. Real time 3D face alignment with Random Forests-based Active Appearance Models. In Proceedings of the IEEE International Conference and Workshops on Automatic Face and Gesture Recognition, Shanghai, China, 22–26 April 2013; pp. 1–8.
51. Sun, J.; Huang, D.; Wang, Y.; Chen, L. A coarse-to-fine approach to robust 3D facial landmarking via curvature analysis and Active Normal Model. In Proceedings of the IEEE International Joint Conference on Biometrics, Clearwater, FL, USA, 29 September–2 October 2014; pp. 1–7.

Article

Towards Real-Time Facial Landmark Detection in Depth Data Using Auxiliary Information

Connah Kendrick [1,*], Kevin Tan [1], Kevin Walker [2] and Moi Hoon Yap [1]

[1] Visual Computing Lab, School of Computing, Mathematics and Digital Technology, Manchester Metropolitan University, Chester Street, Manchester M1 5GD, UK; K.Tan@mmu.ac.uk (K.T.); M.Yap@mmu.ac.uk (M.H.Y.)

[2] Image Metrics Ltd., Manchester M1 3HZ, UK; Kevin.Walker@image-metrics.com

* Correspondence: C.Kendrick@mmu.ac.uk

Received: 15 May 2018; Accepted: 14 June 2018; Published: 17 June 2018

Abstract: Modern facial motion capture systems employ a two-pronged approach for capturing and rendering facial motion. Visual data (2D) is used for tracking the facial features and predicting facial expression, whereas Depth (3D) data is used to build a series of expressions on 3D face models. An issue with modern research approaches is the use of a single data stream that provides little indication of the 3D facial structure. We compare and analyse the performance of Convolutional Neural Networks (CNN) using visual, Depth and merged data to identify facial features in real-time using a Depth sensor. First, we review the facial landmarking algorithms and its datasets for Depth data. We address the limitation of the current datasets by introducing the Kinect One Expression Dataset (KOED). Then, we propose the use of CNNs for the single data stream and merged data streams for facial landmark detection. We contribute to existing work by performing a full evaluation on which streams are the most effective for the field of facial landmarking. Furthermore, we improve upon the existing work by extending neural networks to predict into 3D landmarks in real-time with additional observations on the impact of using 2D landmarks as auxiliary information. We evaluate the performance by using Mean Square Error (MSE) and Mean Average Error (MAE). We observe that the single data stream predicts accurate facial landmarks on Depth data when auxiliary information is used to train the network. The codes and dataset used in this paper will be made available.

Keywords: deep learning; RGB; depth; facial landmarking; merging networks

---

1. Introduction

Motion capture using visual cameras is a common practice in high-end facial animation production. Commercial companies have a preference towards optical marker based systems, such as Vicon [1] as they allow for a large quantity of tracked landmarks with high accuracy. Additionally, with optical markers the addition of multiple cameras allows Depth information to be predicted. However, the set up time of the tracking markers is lengthy and prone to human error. A solution to this is to implement marker-less tracking, which uses visual cameras, computer vision techniques and machine learning to label facial features [2,3]. Marker-less tracking, currently, cannot track as accurately or as many points as optical marker systems. Similarly, to optical markers, additional cameras allow capture of Depth information. However, with technology advancements, the prices of Depth sensors have decreased, while they have significant performance improvements, making them suitable for consumer based production. Additionally, with the availability of RGB with Depth (RGBD) sensors, the potential to increase accuracy is possible by merging the data streams within a neural network. Merging RGB and Depth allows a marker-less system to predict Depth without the requirement of multiple cameras with high accuracy. The Depth information assists greatly in identifying facial feature movement and synthesising to 3D models. Furthermore, in object recognition Greyscale (Gs) outperforms RGB data significantly [4]. Thus we also compare against Gs image and merged Greyscale with Depth (GsD). This study improves upon current work, as the literature is split between networks that use full RGB [2,3] and networks that run Gs [5,6] without justification, and focuses solely on 2D landmarks prediction. 3D landmarks are important for face recognition in the presence of expressions [7] and real-time facial animation [8]. To do this, we extend the existing work to predict 3D landmarks and investigate the impact on 2D and 3D data if they are used as auxiliary information.

As with many fields of research, the implementation of deep learning has shown significant improvements in facial landmarking [9], when compared to traditional machine learning [10]. In this work, we focus on the use of CNNs, like the literature in this area. To perform the experimentation, we develop near identical networks to reduce the deviation between results. Our main contributions are:

- We introduce a new Kinect One [11] dataset, namely KOED to overcome data deficiency in this domain.
- We propose a novel and automated real-time 3D facial landmarks detection method.
- We conduct a complete investigation on the effect of different data streams, such as Gs, RGB, GsD, RGBD in 2D and 3D facial landmarks detection.

By performing this investigation, we can determine the best solution for automated real-time 3D landmarks detection.

2. Related Work

The related work is divided into three sections. Firstly, we give an overview of the current state-of-the-art deep learning to predict facial landmarks. We demonstrate the key aspects of the networks functionality and the features used to localise landmark regions. The second section evaluates merging Gs/RGB and Depth information in a neural network and the current implementation methods. Lastly, we present a review of existing 3D datasets and their limitations.

2.1. Facial Landmarking with Neural Networks

Facial landmarking in deep learning is well established, with state of the art showing both real-time and high accuracy results. Neural networks have solved a wide range of problems, such as facial landmarking, age identification and gender classification. Due to the adaptability of neural networks, previous literature has evolved to use multi-output networks [12,13]. Multi-output networks perform an array of predictions simultaneously, such as age and gender. For our review, we focus on both single and multi-output networks, such as landmark and gender [3] and landmarking only networks. We discuss multiple output networks as they can outperform landmarking only networks as research shows that auxiliary features have a positive effect on network performance [14]. Auxiliary features boost network performance by adding key pieces of information. For example, in age prediction, if gender is used as an auxiliary feature, it aids the network as it learns how the make-up and facial hair affect age prediction. Auxiliary information is predicted by the network in addition to other outputs; the input to the networks is still a single or merged stream of data. Our experiment seeks to observe the effect of different streams of data on a neural network; the area of facial landmarking using auxiliary features, such as age and gender, would be an aspect of future work.

We first evaluate networks that focus solely on the prediction of landmarks. In 2013, Sun et al. [15] proposed an end-to-end network that takes a facial image through a series of convolutions, max-pooling, and fully connected layers, to predict five facial landmarks with reasonable accuracy. Zhou et al. [5] expanded on the work, by proposing a series of detectors to identify facial regions and process them by small neural networks. They also use a refinement approach that aligns the facial features before landmark prediction. Lia et al. [16] proposed a complex network for landmark detection where they implemented a two-stage network, the first stage is a series of convolution and deconvolution layers to process the image given into a high-value feature set. The features were then processed by a series of LSTM [17] layers to identify and refine the landmark position.

Recently, Liu et al. [18] used a multitude of facial feature detectors to identify regions, such as eyes, nose, and mouth. The authors processed these regions with small sized neural networks that identify the landmarks on each of the features. This method achieves high accuracy results, as the network and detectors specialise in different aspects of the face, instead of trying to generalise to all the unique features. However, unlike Zhou et al. [5], they did not align the features.

We now review the work that uses multiple output networks. Zhang et al. [12] experimented in the use of auxiliary features to increase a network understanding of facial structure and features. They created multiple networks with the structure remaining the same except for the outputs changing by adding key pieces of information such as facial direction, age, and gender. By incorporating auxiliary features, networks learned facial features in more Depth. The authors observed a significant increase in accuracy when asking the network to determine these extra features, even when training the network to perform normally difficult tasks, such as facial direction. More recently, Zhang et al. [14] extended their work on facial alignment. Jourabloo et al. [6] used a similar method to predict landmarks by having a series of networks refine the positions. However, they focused on using the landmarks to refine the appearance of a 3D model. Even though Zhang et al. [14] and Jourabloo et al. [6] provide high accuracy networks, the networks require pre-processing to crop faces out of the image.

Finally, we review all-in-one networks, where no pre-processing is required before network prediction. The most recent research for facial landmarking focused on end-to-end networks based upon Recurrent Neural Networks (RNN) [19]. Zhang et al. [2] presented an all-in-one neural network to identify and landmark faces in an image. They used three interlinked networks to refine the landmarking approach. The result of the network is five facial landmarks and bounding box for every face in an image. On the other hand, Ranjan et al. [3] produced their all-in-one network to retrieve the face bounding box, landmark, facial direction and gender with high accuracy. The network included a separate classifier to check if the first section of the network returned a true face.

The networks, when trained on the separate streams of data, give high-end accuracy results starting from the small-scale one output networks to complex multi-model methods. However, the work is limited as it only considers single RGB or Gs images to predict 2D landmarks. Whereas state of art uses multiple cameras or Depth data to estimate the desired 3D landmarks. Additionally, the literature does not give justification for the use of either RGB or Gs. As neural networks are adaptable, we want to investigate how the different streams of data affect a neural network's ability to predict both 2D and 3D landmarks. Furthermore, we extend this by analysing the effect of merging multiple data streams for accurate facial landmark prediction, such as integrating both RGB or Gs with Depth. We also extend on Zhou et al.'s [5] work by analysing the effect of using UV and XYZ as auxiliary features, compared to using UV or XYZ only to train a model that understands facial structure in detail.

Investigation of the use of Depth information to predict facial landmarking has been performed [20]. However, much of the focus is on using surface curvature analysis. Curvature analysis does give reasonable results on low noise models, but it is a slow process and can only track a few points in areas of high curvature change. Another method of predicting 3D facial landmarks is shown by Nair et al. [21], who impressively have predicted a total of 49 landmarks on the face, but they avoid the mouth area. However, this method required a generated 3D model, as point distributed model is used to deform a template face with landmarks assigned to the new mesh. This is an intense and computationally expensive task. Both methods required pre-generated models that are difficult at real-time on a consumer base; our focus is the sole use of images to accurately infer the landmarks.

2.2. Merging Visual and Depth

A multi-model network [22] for the merging of data, such as Gs and Depth, usually implements three separate networks that work together. The first two networks take input from the separate streams of data; then they can be processed the same way as a traditional CNNs. The network uses these convolutions to extract the unique features in each of the data streams. After the processing, the inputs for unique features the outputs are fed into the third neural network and the data merged using basic matrix operations. The third network, similar to the first two networks, functions as a traditional convolution network.

Merging separate streams of data is, in some areas, a common practice, such as in action recognition [23]. Park et al. [23] showed by merging an RGB stream with its optical flow counterpart in a neural network, significantly improves the networks accuracy, by segmenting out the motion in action recognition.

Merging different data streams has also shown increased accuracy in object recognition [25]. Socher et al. [24] use a single layer convolutional neural network to retrieve RGB and Depth images to extract low-level features. The output of these networks is fed into separate RNNs. The results of both RNNs is fed into a softmax classifier. By combing the data, they showed significant improvement in object recognition. The research in this field are inspired by [23,24] on merging data streams to increase the accuracy of detection and recognition systems.

For our experiment, we solve a different type of problem where the detection and recognition system use classification; landmarking is a regression-based problem. Applying classification to a landmarking problem would mean assigning a true or false value for every pixel in an image, which would be too processor intense for real-time performance. Whereas regression allows a single output to be a wide range of values, significantly reducing the processing requirements.

2.3. Existing Datasets

As the experiment required visual and Depth data from the same synchronous capture for both the merging networks and to prevent bias between the RGB only, Gs and Depth only networks, a review of the available datasets was performed. As the result of the neural network is to predict landmark locations in 2D and 3D, the Depth data should be captured from a similar position and angle to the RGB, for near identical recording. As a result of requiring the features to match, datasets that use devices like the Kinect are required, as they use forward facing sensors that are only a few millimetres apart, resulting in similar data view outputs. The available datasets are summarised as follows:

- Face Warehouse [26]: is a large-scale dataset containing 150 participants with an age range of 7–80. The dataset contains RGB images (640 × 480), Depth maps (320 × 240) and 3D models with 74 UV landmarks. The dataset focuses solely on posed expressions giving one model and image when the participant displays the expression. Furthermore, for capture they use the Kinect version 1 [27]. The dataset is captured under different lighting and in different places. As only the expressions peak is captured, there is not a significant amount of data for training deep learning and it is at a low resolution compared to modern cameras. Overall, the Face Warehouse is a good 3D face dataset providing a wide assortment of expressions with landmark annotations, but with no onset or offset of the expression.
- Biwi Kinect Head Pose [28]: is a small-scale Kinect version 1 dataset containing 20 participants, four of the participants were recorded twice. During the recording, keeping a neutral face, the participants would look around the room only moving their heads. The recordings are different lengths. The Depth data has been pre-processed to remove the background of all no face sections. The recording contains no facial landmarks, but the centre of the head and rotation is noted per frame. Although the recording was done in the same environment, the participants can be positioned in different sections of the room changing the background; the lighting remains consistent. Overall, the Biwi Kinect dataset was not suitable for the experiment as it contained no facial expressions and was recorded using the Kinect version 1.
- Eurocom Kinect [29]: is a medium-sized dataset containing 52 participants, each participant was recorded twice with around two weeks in between. Participants were recorded by having single images of them performing nine different expressions. The images were taken using the Kinect version 1 and images were pre-processed to segment the heads. The coordinates for the cropping are given as well as six facial landmarks. The Eurocom dataset contains few images

for a deep learning network and is recorded with the Kinect version 1, making it unsuitable for the experiment.

- VAP face database [30]: is a small size dataset containing 31 participants. The dataset was recorded using an updated Kinect version 1 for Windows, this version gives a bigger RGB image (1280 × 1024) and larger Depth map (640 × 480), but at the cost of reduced frame rates. The recording was also done using the Kinects 'near-mode' which allows for the increased resolution described. Each participant has 51 images of the face taken at different head angles performing a neutral face and some frontal face with expressions. The recordings were done in the same place with consistent lighting. As the dataset contains single images and few participants performing facial expressions, it is unsuitable for the experiment, but for head pose estimation it would be appropriate.
- 3D Mask Attack [31]: is a small to medium scale dataset containing 17 participants, but a large collection of recordings. The participant is recorded in three different sessions; in each session the participant is recorded five times for 300 frames per recording, holding a neutral expression. The recording uses the Kinect version 1. The eyes are annotated every 60 frames with interpolation for the other frames. The recordings were done under consistent lighting and background. The 3D Mask Attack dataset contains a vast number of frames, but all use the neutral expression, face the camera and use the older Kinect making it unsuitable for the experiment.

The existing datasets do not meet the following requirements:

- Deep learning requires large-scale datasets containing many thousands of training examples.
- Facial expression is key for robust landmarking systems, including the onset and offset of expressions.
- Facial Landmarks, in both 2D and 3D.
- As facial movement can be subtle, high-resolution images are required, which is why Kinect version 2 with both higher accuracy and resolution is needed.
- Real-time frame rates, as most systems target 30 Frames Per Second (FPS).

3. Proposed Method

3.1. Kinect One Expression Dataset (KOED)

As currently available datasets did not meet the requirements of the project, we created an in-house dataset. All networks were trained using the in-house dataset.

3.1.1. Experimental Protocol

The experiment comprised of replicating seven universal expressions. Participants were instructed to begin with a neutral face, perform the expression and then return to the neutral face. We also record a full clip of the participant performing a neutral expression. The expressions performed are as follow:

- Happy
- Sad
- Surprise
- Anger
- Fear
- Contempt
- Disgust

All participants volunteered for the experiment with no monetary reward. To obtain a wide range of diversity, anyone over the age of 18 was able to join the experiment. The dataset has 35 participants, with a wide range of ages. The majority of the clips are female with a majority of white British, but it does include participants from Saudi Arabia, India and Malaysia.

3.1.2. Emotional Replication Training

During each recording, a trained individual was present to advise the participants on facial expressions, providing some prior training. However, during the recording the trainer would not give any advice to prevent distraction.

3.1.3. Ethics

Ethics was reviewed and approved by the Manchester Metropolitan University ethics committee (SE151621).

3.1.4. Equipment and Experimental Set up

The experiment was set up in the same room for each participant to ensure each recording was done similarly. We used a green screen recording room for each of the recordings; this allowed a consistent background and lighting. The participant sat in the centre of the room, where the lights could be placed at even distances to ensure consistent coverage. The studio has six lights that were evenly spaced around the participant, in a backward C shape; we used a series of back-lights to ensure the background was also lit up. The Kinect was placed one meter away from the participants, at their head height while they sat down. Steps were taken to ensure consistent lighting, but to ensure ground truth colour was available we use a colour checker placed to the left of the participants. The participant was required to remain still during the recording. As recording both RGB and Depth requires a large quantity of data to be stored, we used a SSD fitted laptop. An example of the experimental set up is shown in Figure 1.

Figure 1. An example of the data capture set up.

3.1.5. Camera

The camera used was the Kinect for Xbox One, which gives synchronous streams of both RGB and Depth data at 30 fps. As the Kinect performs better after reaching working temperature, we turn the sensor on 25 min prior to any recording to ensure high quality data capture.

3.1.6. Lighting

We use six ARRI L5-c LED directional lights focusing on the individual participant. The lights are set to emit white light only to prevent any discoloring of the participants faces. The backlighting is done with a series of photo beard tungsten fluorescent tubes.

3.1.7. Frame Rate and Storage

We record at the Kinect's maximum capabilities, RGB (1920 × 1080) and Depth (512 × 424) at 30 fps, for speed we save both files in binary format. The images stored are unmodified from the ones received from the Kinect, no lossy compression is implemented. As the data is stored in raw binary format the dataset requires, at the time of writing, over 675 GB of storage for the full dataset.

3.2. Methodology

We implement multiple near-identical networks that function by pre-processing the image with convolutions with Rectified Linear Unit (ReLU) activations and then a series of fully connected layers to determine the final output. We illustrate the base networks in Figure 2. The base networks take a single stream of data, Gs, RGB or Depth and process through a series of convolutions to extract facial features. We use max-pooling to focus on high level features, and decrease processing requirements, but take into consideration that this can negatively impact accuracy [32]. The network utilises ReLU as an activation function after each convolutional layer as it does not normalise data. The resulting feature maps are then processed by fully connected layers to predict the facial landmarks. For the second stage, we examine the effectiveness of merging data streams, RGBD and GsD, we have a multiple input model, shown in Figure 3. The merge network used two CNNs: one to take the RGB/ Gs image; and another to take the Depth image. The two networks then use a series of convolutions to extract unique features from each of the inputs. The results of the two CNNs are combined and used as input to a third network. The third network further convolutes over the images giving a high value feature set for the fully connected layer.

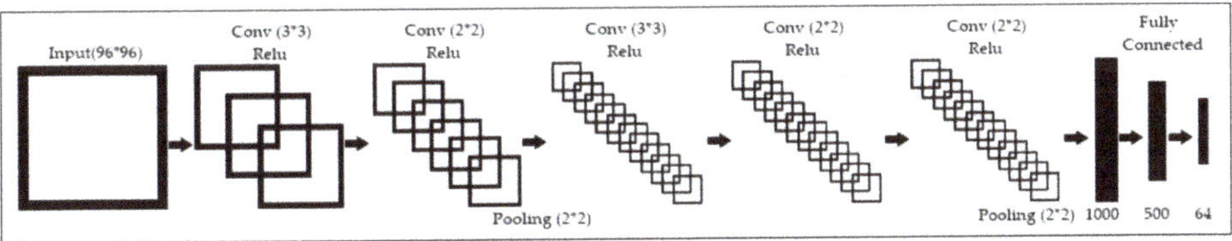

Figure 2. A visualisation of the basic network used for this experiment.

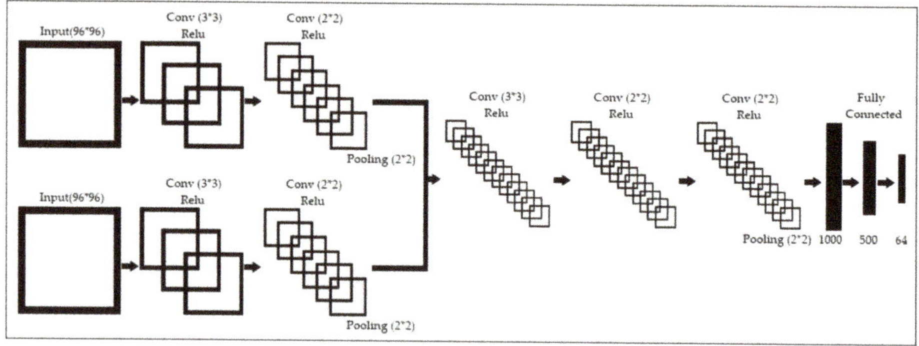

Figure 3. A visualisation of the merged network used for this experiment.

As auxiliary features do affect how the network learns and XYZ points are desired, but not commonly predicted, we repeat the experiments not just with different data streams, but alternative outputs. The different outputs aid in showing how the networks can understand and learn both the features and facial structure, in different spaces. The three types of outputs and their metrics that we train the networks to predict are:

- The UV coordinates, in pixels
- The XYZ coordinates, in meters
- The UVXYZ coordinates

Where the UV points are the 2D image landmark coordinates and the XYZ points are the 3D location of the landmarks in camera space. As the outputs are in non-compatible metrics, they cannot be predicted in the same fully connect layer. To overcome this, we propose a multi-model output, where the final convolutional outputs are fed into different output models. This means, for UV and XYZ, there will be one model of fully connected layers for the convolution to be passed into. However, the UVXYZ network will have the convolutions output into two different models, one for UV calculation and one for XYZ. Traditionally with the Kinect, we require the 2D landmarks and use them to reconstruct the 3D points with a Depth map. Furthermore, by asking a network to infer UV and XYZ points, it could adopt the similar methodology, thus improving performance.

The networks are trained with a batch size of 240 using a stride of one over 100 epochs, using tensor-flow [33] with the Keras [34] API. We used the KOED dataset with 10-fold cross-validation; this ensures the network is trained, validated and tested on multiple participants, illustrating reliability. The cross-validation split was performed semi-randomly, with 70% training, 20% validation and 10% testing, ensuring no participant existed in multiple sets. We use MSE as our loss function, shown in Equation (??), using Adam [35] as the optimiser. MSE has more emphasises on large numbers allowing for large outliers to be resolved during training. However, we also calculate the MAE, as shown in Equation (??). MAE gives equal weight to all the errors illustrating the overall error. By using these error functions, we can determine the number of errors the networks produce and the size of errors. We use MSE for training as it is traditional in regression-based deep learning.

$$\text{MSE} = \sum_{i=0}^{n} \frac{(y_i - y_i')}{n} \tag{1}$$

where:

- n is the number of samples in the training batches.
- y_i is the ground truth for the training image.

- y_i' is the predicted output for the training image.

$$\mathrm{MAE} = \sum_{i=0}^{n} \frac{|y_i - y_i'|}{n} \tag{2}$$

where:

- n is the number of samples in the training batches.
- y_i is the ground truth for the training image.
- y_i' is the predicted output for the training image.

4. Results

To compare the networks, we first show the validation during training and examine the performance of each stream. For each of the results we start with the UV (2D), then XYZ (3D) and finally, the UV XYZ (All) results. After this, we show an evaluation of the networks on testing data and the feature maps produced by the networks. Finally, we examine the results of the testing set with both MSE and MAE scores.

Figure 4 illustrates that for the prediction of UV landmarks, both RGB and Gs converge at similar epochs, 40. In addition, they both share many similar traits, such as that they both start with a significantly lower loss and have more stable learning than input streams that incorporate Depth. Overall, RGB performs the best in both MSE and MAE. The networks that merge visual and Depth data converge much later than RGB and Gs, but their results of MSE are close to the RGB and Gs scores. RGBD and GsD have unstable learning curves and encounter hidden gradients that cause loss to increase rapidly. The single channel GsD converges earlier than RGBD, indicating that a single clean frame learns faster on how to smooth a noisy Depth map than a three channel RGB image. The single channel Depth encounters the most unstable learning and converges at a much later stage, showing without a visual stream to assist the Depth data cannot easily locate UV landmarks. Furthermore, this is illustrated by Depth performing the worst when evaluated on MSE and MAE.

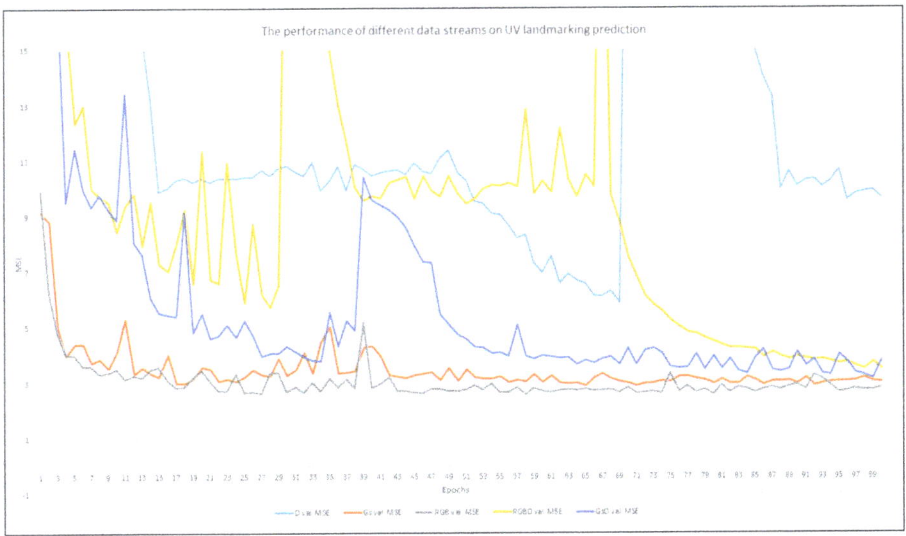

Figure 4. The MSE of the UV Only networks validation over 100 epochs.

Figure 5 illustrates the MSE of the XYZ only network, like UV, RGB and Gs start with a low loss and converge the quickest at around epoch 30. However, the learning is unstable, indicating retrieving accurate 3D landmarks from visual images is a difficult task, although in the final epoch RGB has the lowest MSE. The input streams that incorporate Depth converge sooner than in the UV prediction networks. Furthermore, their learning rate is more stable than the RGB and Gs stream, but hidden gradients are still an issue. In addition, they converge at a similar location slightly higher than RGB and Gs, although at some point they score lower loss than the RGB and Gs networks. This convergence also occurs after a hidden gradient, indicating there is a shared local minimum caused by the inclusion of Depth data, the most prominent of these is GsD, which consistently has the lowest loss over epochs until it reaches a hidden gradient, to which it then becomes the worst performing stream.

Figure 6 illustrates the MSE of the UVXYZ networks, where RGB and Gs begin with the lowest loss, but RGB has a significantly lower loss than Gs. The learning rates of RGB and Gs are stable and converge quickly around epoch 43, with Gs performing the best. The input streams that incorporate Depth data also converge quickly, with Depth and GsD having stable learning rates, unlike RGBD. Furthermore, hidden gradients are still an issue. However, unlike in UV and XYZ only networks, the UVXYZ quickly recovers. This demonstrates how auxiliary information is benefiting the networks ability to learn from the different data streams by overcoming issues, such as the local minimum seen in Figure 5.

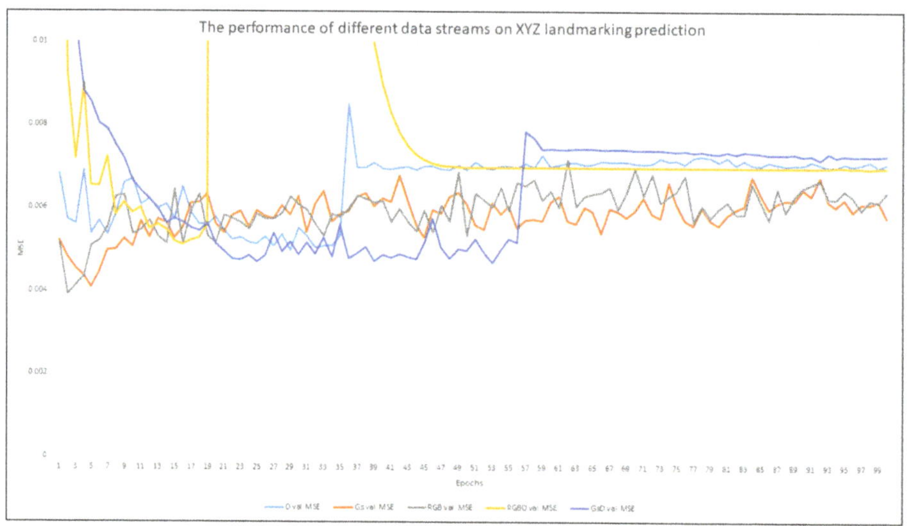

Figure 5. The MSE of the XYZ Only networks validation over 100 epochs.

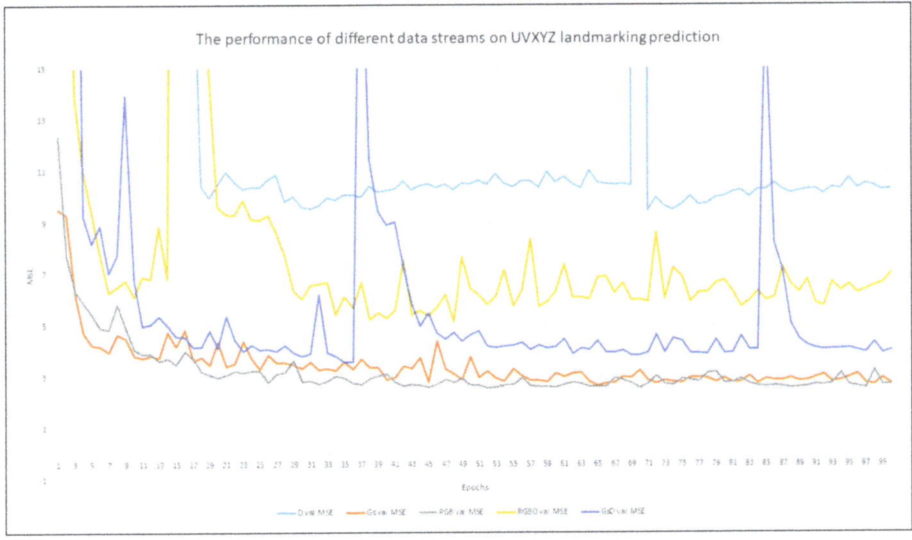

Figure 6. The MSE of the UVXYZ networks validation over 100 epochs.

Figure 7 visually compares the results in both 2D and 3D. We summarise the observations:

- In the UV only prediction, the results are visually similar, but there is some deviation between each of the networks. When using Depth as the input stream, the predictions of both the right eye and lip corners are predicted less precise than the other input streams; this could be directly affected by the noise in the Depth maps, as when merged with a visual stream, performance is improved.
- For UVXYZ, there is no noticeable difference between the UV results.
- For the XYZ only predictions we see much larger discriminations in the predicted facial landmarks. Some of the major changes are:
  - From the frontal view there is a variation in the mouth width, with Gs being the smallest and Depth being the widest.
  - Nose landmarks shifts in GsD were the nose tip and right nostril are predicted close to each other.
  - Eye shape changes between networks, Gs and RGBD produce round smooth eyes. Whereas others are more jagged and uneven.
  - From the side view, we see the profile of the face change with the forehead and nose shape varying greatly between networks.
- In contrast to the UV results in the UVXYZ network, with the addition of auxiliary information the resulting geometric landmarks on the mouth, nose, eye and eyebrows, become more precise and consistent. In most of the cases the eyes are smoother, the eyebrows are more evenly spaced, the nose irregularity in GsD no longer occurs and the mouth width consistency has improved greatly. These results show that, as UV is easier for the networks to learn as all streams manage similar results, when used as auxiliary information, they aid to standardise the 3D locations as well. However, there are still some variations in the profile of the nose and in RGB the right eye is predicted to be shut.

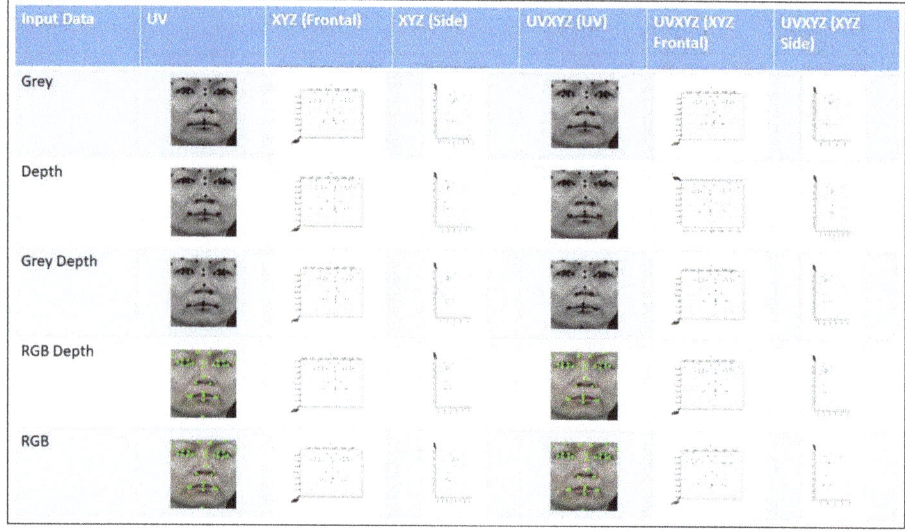

Figure 7. A visual comparison of the results from the trained networks.

As shown in Table 1, for UV landmarks RGB has the lowest MSE, with Gs not far behind. It also shows that for predicting landmarks in 3D only, that having both a visual and Depth data allows for the highest precision results, with RGBD and GsD scoring the lowest with marginal differences in score. For the MAE and MSE of the UVXYZ networks, we show the separate stages of the loss calculation:

- Combined loss, which is the sum of UV and XYZ layers loss.
- UV loss, the loss of the UV layers alone.
- XYZ loss, the loss of the XYZ layers alone.

The combined loss shows the overall network performance, but the UV and XYZ alone show the networks' performance on the individual outputs. By comparing the loss of the UV and XYZ alone, we illustrate how the auxiliary information is affecting network performance, compared to networks predicting UV only or XYZ only landmarks. When trying to predict UVXYZ data, Gs performs the best overall. We show that by introducing the 3D landmarks, we reduce the overall loss significantly to UV alone in both RGB, Gs and GsD networks. Furthermore, the prediction of XYZ is improved in the same networks. We see similar results in the MAE, shown in Table 2, where networks reduce the loss below the UV alone networks. However, RGB sees the least MAE for UV. For overall combined loss and XYZ loss, Gs scores the lowest in MSE and MAE.

Table 1. Table of the testing set evaluation on MSE. Bold highlights the lowest error.

Input Data	UV MSE	XYZ MSE	UVXYZ MSE (Combined)	UVXYZ MSE (UV)	UVXYZ MSE (XYZ)
Gs	1.8192	0.0023	**1.3695**	**1.3676**	**0.0019**
Depth	6.4672	0.0023	6.6509	6.6482	0.0027
Gs Depth	2.1845	0.0022	1.8933	1.8911	0.0022
RGB Depth	2.1561	**0.0022**	2.8744	2.8752	0.0022
RGB	**1.7488**	0.0023	1.5612	1.5592	0.0019

Table 2. Table of the testing set evaluation on MAE. Bold highlights the lowest error.

Input Data	UV MAE	XYZ MAE	UVXYZ MAE (Combined)	UVXYZ MAE (UV)	UVXYZ MAE (XYZ)
Gs	1.0052	**0.0341**	**0.9127**	**0.8797**	0.0330
Depth	1.9150	0.0361	1.9705	1.9322	0.0382
Gs Depth	1.1210	0.0379	1.0617	1.0246	0.0371
RGB Depth	1.0848	0.0367	1.3056	1.2685	0.0371
RGB	**0.9553**	0.0346	0.9685	0.9388	**0.0297**

The key differences in single task networks and multi-task networks in predicting facial landmarks were observed in the feature maps of the networks, illustrated in Figure 8. The network kernels learned the spatial information from UV prediction. Therefore, the feature maps shown in the UV prediction demonstrate the activation of appearance-based facial features. On the other hand, when predicting the geometry coordinate of XYZ, we observed that the feature maps of the convolutional layers had point-based (facial landmarks) activation. This is due to the Z component which makes the facial landmarks more separable. The UVXYZ column depicts the features maps in UVXYZ prediction. We observed it has better pattern representation with both appearance based and point/landmarks information. The Gs network performs the best with the feature maps demonstrating the networks can process the input stream to focus on the specific landmark regions of the face. Further advantages occur when auxiliary information is added: the kernels become refined and are able to detect features with high intensity, as the network is forced to learn how the structure appears in both 2D and 3D. It also means the network can process the data more efficiently as the input is a single stream. However, a disadvantage of this system is that the image must be pre-processed from RGB to Gs.

To demonstrate the effectiveness of the network, we visualise the predicted landmarks of the Gs network on a 3D model, shown in Figure 9 (see Supplementary Materials). With Gs as input data stream, our proposed method predicts accurate 3D facial landmarks on raw Depth data using auxiliary information. Furthermore, this illustrates the accuracy of the network, even with raw Depth data, our proposed method manages to estimate accurate 3D facial landmarks after pre-processing to crop and resize Depth images for the network, where a human would be incapable of without full-size Depth images [36]. However, due to the noise from the raw data, the limitation of our proposed method is not able to locate the Z position precisely in some cases.

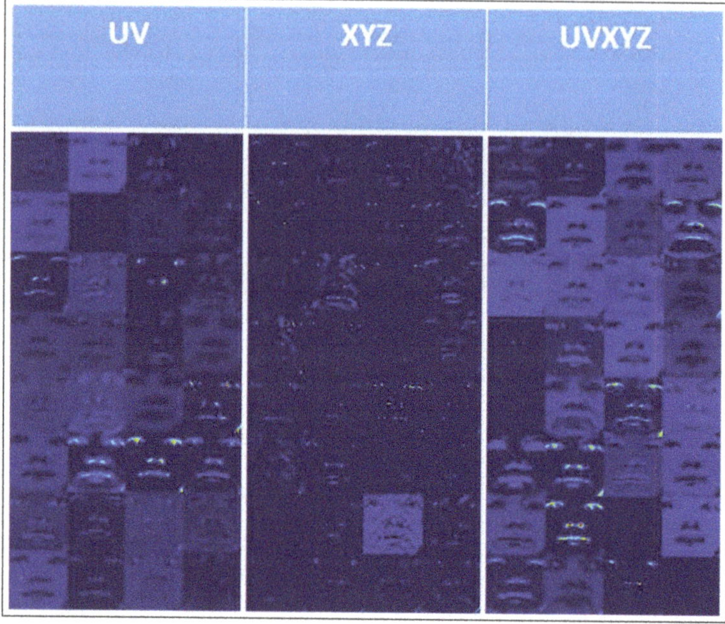

Figure 8. A comparison of the output of the final convolutional filter for each type of network prediction on the RGB Images. The third column illustrates the feature maps for UVXYZ prediction, the best performance with auxiliary information.

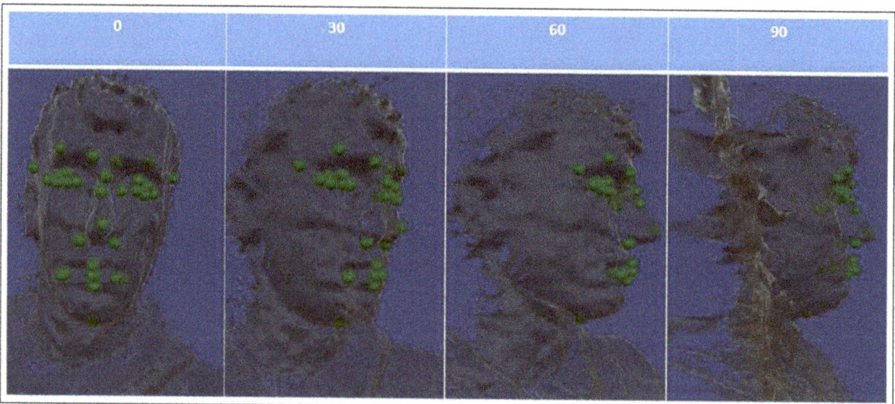

Figure 9. The result of the Gs UVXYZ trained network and the appropriate model from the same input Depth map. The model is transparent to show geometry coordinates of the facial landmarks.

5. Discussion and Conclusions

In this work we have shown and illustrated the effect of different data streams within neural networks, to identify which streams are ideal for current research topics, as current literature uses a mixture. We also extended the work by the prediction of points in the camera (XYZ) space as this is a valuable resource in facial expression recognition and animation synthesis, but current literature focuses on image (UV) space coordinate systems. Unique insights into each stream of data were obtained, demonstrating the pros and cons of each stream. To prevent bias, an in-house dataset was used, showing that each network could reliably track facial features and expressions in both 2D and 3D. The networks showed that the existing data-streams could accurately predict 2D and 3D landmarks.

Comparing the results and feature maps of the networks demonstrates the ability of the networks to process and understand the different forms of data and if they are beneficial to the network. Full RGB performed the most effectively on UV with the least amount of errors and the lowest scale of errors. While Depth shows its effectiveness at predicting landmarks, the noise it presents requires additional streams, such as RGB to smooth out and retrieve reliable results. In the final experiment, for predicting UVXYZ, we show that although for UV alone RGB is the most efficient, Gs outperformed it, illustrating that more generalizable single frames are more effective when predicting a wide range of values. While Depth has shown to be difficult for the networks to learn from, with limitations such as exploding gradients, even after merging with cleaner streams it has been shown to be effective even when cropped and resized for the prediction of landmarks, where traditional methods require full-size Depth images.

This work focused exclusively on the use of neural networks to predict facial landmarks without the aid of physical markers, sensors, or reference points placed on the individuals. There have been many incremental studies into the use of neural networks to predict the image (UV) space landmarks successfully. However, the results all use different streams of data with little consensus on why the stream is used, except for dataset or memory limitations. In addition, XYZ coordinates are not being predicted by neural networks in current systems. For networks, many industries desire the use of 3D landmarks in real-time.

There are several limitations in this study, mostly related to the data used to train the network and the difficulty of 3D landmarks. Firstly, due to the context issue of cropping, a Depth map recording was done in a controlled environment, so the network must only learn a manageable part of the 3D viewing frustum. This, regarding animation, has an advantage as it normalises the facial position, while still tracking 3D facial movement. However, for full 3D prediction full Depth maps would still be required. Future work should seek out new technologies, such as the Intel real-sense [37], which could resolve the noise issue of the Kinect as it provides both higher resolution and cleaner Depth maps as shown by Carfagni et al. [38], which would aid the networks' ability to learn from the data. Other aspects would be to further the work with a larger dataset to test the reliability of no Depth streams with a wider demographic of faces.

We have shown and analysed how the input data stream can affect a deep neural network framework, for the analysis of facial features, which can have an impact on facial recognition, reconstruction, animation, and security, by providing how the networks interact with the different data streams. The stream shows different levels of accuracy and reliability which can positively affect future work. Future work will include increasing the number of participants and increasing the amount of reliably tracked landmarks without marker 3D reference points on the face, as current literature is limited in this area.

6. Materials and Methods

We provide access to all codes used to build and train models on GitHub. We also provide demo codes to enable the real-time use of the trained models, with the use of a Kinect. All scripts are provided in python. The in-house KOED dataset will be made publicly available. However, in its raw form, the dataset requires over 675 GB to store at the time of writing, without any annotations.

Supplementary Materials: We provide multiple videos representing our results. Firstly, we provide a video of the model and points shown in Figure 9, rotating between ±90 degrees, as it is a raw Depth map model there is no back, thus 360 provides no additional information. Finally, we provide videos demonstrating the feature maps of the networks to illustrate which features in the images the network deems most valuable to the prediction.

Author Contributions: C.K. and M.H.Y. designed the experiment. C.K. performed the experiments. C.K., K.T., K.W., and M.H.Y. analysed the data. All authors were involved in writing the paper.

Acknowledgments: The authors would like to thank their funders: Manchester Metropolitan University Faculty of Science and Engineering (Studentship number: 12102083) and The UK Royal Society Industry Fellowship (IF160006). We received funding to cover the cost of publishing in open access.

Conflicts of Interest: The authors declare no conflict of interest.

Abbreviations

The following abbreviations are used in this manuscript:

CNN	Convolutional Neural Network
LSTM	Long Short-Term Memory
RNN	Recurrent Neural Network
ReLU	Rectified Linear Unit
RGB	Red Blue Green
RGBD	Red Blue Green Depth
Gs	Greyscale
GsD	Greyscale Depth
D	Depth
2D	Two Dimensional
3D	Three Dimensional
KOED	Kinect One Expressional Dataset
HD	High Definition
MSE	Mean Squared Error
MAE	Mean Absolute Error

References

1. Vicon Motion Systems Ltd. *Capture Systems*; Vicon Motion Systems Ltd.: Oxford, UK, 2016.
2. Zhang, K.; Zhang, Z.; Li, Z.; Qiao, Y. Joint Face Detection and Alignment Using Multitask Cascaded Convolutional Networks. *IEEE Signal Process. Lett.* **2016**, *23*, 1499–1503. [CrossRef]
3. Ranjan, R.; Patel, V.M.; Chellappa, R. HyperFace: A Deep Multi-task Learning Framework for Face Detection, Landmark Localization, Pose Estimation, and Gender Recognition. *IEEE Trans. Pattern Anal. Mach. Intell.* **2016**, 1. [CrossRef]
4. Bui, H.M.; Lech, M.; Cheng, E.; Neville, K.; Burnett, I.S. Using grayscale images for object recognition with convolutional-recursive neural network. In Proceedings of the 2016 IEEE Sixth International Conference on Communications and Electronics (ICCE), Ha Long, Vietnam, 27–29 July 2016; pp. 321–325. [CrossRef]
5. Zhou, E.; Fan, H.; Cao, Z.; Jiang, Y.; Yin, Q. Extensive facial landmark localization with coarse-to-fine convolutional network cascade. In Proceedings of the 2013 IEEE International Conference on Computer Vision Workshops, Sydney, NSW, Australia, 2–8 December 2013; pp. 386–391. [CrossRef]
6. Jourabloo, A.; Liu, X. Large-Pose Face Alignment via CNN-Based Dense 3D Model Fitting. In Proceedings of the 2016 IEEE Conference on Computer Vision and Pattern Recognition (CVPR), Las Vegas, NV, USA, 27–30 June 2016; pp. 4188–4196. [CrossRef]
7. Han, X.; Yap, M.H.; Palmer, I. Face recognition in the presence of expressions. *J. Softw. Eng. Appl.* **2012**, *5*, 321. [CrossRef]
8. Faceware Technologies Inc. *Faceware*; Faceware Technologies Inc: Sherman Oaks, CA, USA, 2015.
9. Feng, Z.H.; Kittler, J. Advances in facial landmark detection. *Biom. Technol. Today* **2018**, *2018*, 8–11. [CrossRef]
10. Phillips, P.J.; Flynn, P.J.; Scruggs, T.; Bowyer, K.W.; Chang, J.; Hoffman, K.; Marques, J.; Min, J.; Worek, W. Overview of the Face Recognition Grand Challenge. In Proceedings of the 2005 IEEE Computer Society Conference on Computer Vision and Pattern Recognition (CVPR'05), San Diego, CA, USA, 20–25 June 2005; IEEE Computer Society: Washington, DC, USA, 2005; Volume 1, pp. 947–954. [CrossRef]
11. Microsoft. *Microsoft Kinect*; Microsoft: Redmond, WA, USA, 2013.

12. Zhang, Z.; Luo, P.; Loy, C.C.; Tang, X. Facial landmark detection by deep multi-task learning. In *Lecture Notes in Computer Science*; (Including Subseries Lecture Notes in Artificial Intelligence and Lecture Notes in Bioinformatics); Springer: Berlin, Germany, 2014; Volume 8694, pp. 94–108.
13. Hand, E.M.; Chellappa, R. Attributes for Improved Attributes: A Multi-Task Network for Attribute Classification. *arXiv*, **2016**, arXiv:1604.07360.
14. Zhang, Z.; Luo, P.; Loy, C.C.; Tang, X. Learning Deep Representation for Face Alignment with Auxiliary Attributes. *IEEE Trans. Pattern Anal. Mach. Intell.* **2016**, *38*, 918–930. [CrossRef] [PubMed]
15. Sun, Y.; Wang, X.; Tang, X. Deep convolutional network cascade for facial point detection. In Proceedings of the IEEE Computer Society Conference on Computer Vision and Pattern Recognition, Portland, OR, USA, 23–28 June 2013; pp. 3476–3483. [CrossRef]
16. Lai, H.; Xiao, S.; Pan, Y.; Cui, Z.; Feng, J.; Xu, C.; Yin, J.; Yan, S. Deep Recurrent Regression for Facial Landmark Detection. *arXiv*, **2015**, arXiv:1510.09083. [CrossRef]
17. Hochreiter, S.; Schmidhuber, J. Long short-term memory. *Neural Comput.* **1997**, *9*, 1735–1780. [CrossRef] [PubMed]
18. Liu, H.; Lu, J.; Feng, J.; Zhou, J. Learning Deep Sharable and Structural Detectors for Face Alignment. *IEEE Trans. Image Process.* **2017**, *26*, 1666–1678. [CrossRef] [PubMed]
19. Angeline, P.J.; Angeline, P.J.; Saunders, G.M.; Saunders, G.M.; Pollack, J.B.; Pollack, J.B. An evolutionary algorithm that constructs recurrent neural networks. *IEEE Trans. Neural Netw.* **1994**, *5*, 54–65. [CrossRef] [PubMed]
20. Dibeklioglu, H.; Salah, A.A.; Akarun, L. 3D Facial Landmarking under Expression, Pose, and Occlusion Variations. In Proceedings of the 2008 IEEE Second International Conference on Biometrics: Theory, Applications and Systems, Arlington, VA, USA, 29 September–1 October 2008; pp. 3–8.
21. Nair, P.; Cavallaro, A. 3-D Face Detection, Landmark Localization, and Registration Using a Point Distribution Model. *IEEE Trans. Multimed.* **2009**, *11*, 611–623. [CrossRef]
22. Ngiam, J.; Khosla, A.; Kim, M.; Nam, J.; Lee, H.; Ng, A.Y. Multimodal Deep Learning. In Proceedings of the 28th International Conference on Machine Learning (ICML), Orlando, FL, USA, 3–7 November 2014; pp. 689–696. [CrossRef]
23. Park, E.; Han, X.; Tamara, L.; Berg, A.C. Combining Multiple Sources of Knowledege in Deep CNNs for Action Recognition. In Proceedings of the 2016 IEEE Winter Conference on Applications of Computer Vision (WACV), Lake Placid, NY, USA, 7–9 March 2016; pp. 1–8.
24. Socher, R.; Huval, B. Convolutional-recursive deep learning for 3D object classification. In *Advances in Neural Information Processing Systems 9: Proceedings of the 1996 Conference*; MIT Press Ltd.: Cambridge, MA, USA, 2012; pp. 1–9.
25. Liu, L.; Shao, L. Learning discriminative representations from RGB-D video data. *IJCAI* **2013**, *1*, 1493.
26. Cao, C.; Weng, Y.; Zhou, S.; Tong, Y.; Zhou, K. FaceWarehouse: A 3D facial expression database for visual computing. *IEEE Trans. Vis. Comput. Graph.* **2014**, *20*, 413–425. [CrossRef] [PubMed]
27. Microsoft. *Microsoft Kinect 360*; Microsoft: Redmond, WA, USA, 2010.
28. Fanelli, G.; Dantone, M.; Gall, J.; Fossati, A.; Van Gool, L. Random Forests for Real Time 3D Face Analysis. *Int. J. Comput. Vis.* **2013**, *101*, 437–458. [CrossRef]
29. Min, R.; Kose, N.; Dugelay, J.L. KinectFaceDB: A Kinect Face Database for Face Recognition. *IEEE Trans. Syst. Man Cybern. A* **2014**, *44*, 1534–1548. [CrossRef]
30. Hg, R.I.; Jasek, P.; Rofidal, C.; Nasrollahi, K.; Moeslund, T.B.; Tranchet, G. An RGB-D database using microsoft's kinect for windows for face detection. In Proceedings of the 2012 8th International Conference on Signal Image Technology and Internet Based Systems, (SITIS'2012), Naples, Italy, 25–29 November 2012; pp. 42–46. [CrossRef]
31. Erdogmus, N.; Marcel, S. Spoofing in 2D face recognition with 3D masks and anti-spoofing with Kinect. In Proceedings of the 2013 IEEE 6th International Conference on Biometrics: Theory, Applications and Systems (BTAS), Arlington, VA, USA, 29 September–2 October 2013. [CrossRef]
32. Kendrick, C.; Tan, K.; Walker, K.; Yap, M.H. The Application of Neural Networks for Facial Landmarking on Mobile Devices. In Proceedings of the 13th International Joint Conference on Computer Vision, Imaging and Computer Graphics Theory and Applications (VISAPP), Funchal, Portugal, 27–29 January 2018; INSTICC/SciTePress: Setúbal, Portugal, 2018; Volume 4, pp. 189–197. [CrossRef]

33. Abadi, M.; Barham, P.; Chen, J.; Chen, Z.; Davis, A.; Dean, J.; Devin, M.; Ghemawat, S.; Irving, G.; Isard, M.; et al. TensorFlow: A System for Large-Scale Machine Learning. *Osdi* **2016**, *16*, 265–283.
34. Chollet, F. Keras. 2016. Available online: https://keras.io/ (accessed on 15 June 2018).
35. Kingma, D.P.; Ba, J.L. Adam: A Method for Stochastic Optimization. *arXiv*, **2015**, arXiv:1412.6980v8.
36. Kendrick, C.; Tan, K.; Williams, T.; Yap, M.H. An Online Tool for the Annotation of 3D Models. In Proceedings of the 2017 12th IEEE International Conference on Automatic Face & Gesture Recognition (FG 2017), Washington, DC, USA, 30 May–3 June 2017; pp. 362–369. [CrossRef]
37. Intel. *RealSense SR300*; Intel: Santa Clara, CA, USA, 2016.
38. Carfagni, M.; Furferi, R.; Governi, L.; Servi, M.; Uccheddu, F.; Volpe, Y. On the Performance of the Intel SR300 Depth Camera: Metrological and Critical Characterization. *IEEE Sens. J.* **2017**, *17*, 4508–4519. [CrossRef]

MDPI
St. Alban-Anlage 66
4052 Basel
Switzerland
Tel. +41 61 683 77 34
Fax +41 61 302 89 18
www.mdpi.com

Symmetry Editorial Office
E-mail: symmetry@mdpi.com
www.mdpi.com/journal/symmetry

www.ingramcontent.com/pod-product-compliance
Lightning Source LLC
LaVergne TN
LVHW070616170726
843515LV00004B/97

* 9 7 8 3 0 3 9 3 6 9 6 4 5 *